ESSENTIALS OF FOOD SAFETY AND SANITATION
SECOND EDITION

David McSwane, H.S.D.

Nancy Rue, Ph.D.

Richard Linton, Ph.D.

Prentice Hall
Upper Saddle River, NJ 07458

Library of Congress Cataloging-in-Publication Data

McSwane, David Zachary.
   Essentials of food safety and sanitation / David McSwane, Nancy Rue, Richard
Linton.--2nd ed.
     p. cm.
   Includes bibliographical references and index.
   ISBN 0-13-017371-1
   1. Food service--Sanitation. 2. Food service--Safety measures. 3. Food
handling--Safety measures. I. Rue, Nancy. II. Linton, Richard. III. Title.

TX911.3.S3 M38 2000
363.72'96--dc21

                                           99-053961

Acquisitions Editor: *Neil Marquardt*
Production Editor: *Barbara Cappuccio*
Managing Editor: *Mary Carnis*
Camera-ready Copy Editor: *Karen J. Hall*
Development Editor: *Anna Graf Williams, Ph.D.*
Director of Manufacturing and Production: *Bruce Johnson*
Manufacturing Buyer: *Ed O'Dougherty*
Marketing Manager: *Shannon Simonson*
Editorial Assistant: *Susan Kegler*
Printer/Binder: *R.R. Donnelley & Sons*
Cover Design: *Amy Rosen*
Cover Illustration: *John Wise*

©2000, 1998 by Prentice-Hall, Inc.
Upper Saddle River, New Jersey 07458

Printed in the United States of America

10   9 8 7 6 5 4 3 2 1

ISBN 0-13-017371-1

Prentice-Hall International (UK) Limited, *London*
Prentice-Hall of Australia Pty. Limited, *Sydney*
Prentice-Hall Canada Inc., *Toronto*
Prentice-Hall Hispanoamericana, S.A., *Mexico*
Prentice-Hall of India Private Limited, *New Delhi*
Prentice-Hall of Japan, Inc., *Tokyo*
Prentice-Hall Singapore Pte. Ltd.
Editora Prentice-Hall do Brasil, Ltda., *Rio de Janerio*

CONTENTS

CHAPTER 2—HAZARDS TO FOOD SAFETY   25

CHAPTER 3—FACTORS THAT AFFECT FOODBORNE ILLNESS

CHAPTER 4—FOLLOWING THE FOOD PRODUCT FLOW

Contents

CHAPTER 6—FACILITIES, EQUIPMENT, AND UTENSILS

CHAPTER 7—CLEANING AND SANITIZING OPERATIONS   249

CHAPTER 8—ENVIRONMENTAL SANITATION AND MAINTENANCE

CHAPTER 9—ACCIDENT PREVENTION AND CRISIS MANAGEMENT

CHAPTER   11—FOOD SAFETY REGULATIONS   *355*

PREFACE

Food safety has become an issue of special importance for the retail food industry. There are many opportunities for food to be contaminated between production and consumption. Food can be contaminated at the farm, ranch, orchard, or in the sea. Food also can be contaminated at food processing plants and during transport to food establishments. Finally, food can be contaminated during the last stages of production, at retail establishments, and by consumers in their homes.

Food safety is especially critical in retail food establishments because this may be the last opportunity to control or eliminate the hazards that might contaminate food and cause foodborne illnesses. Even when purchased from inspected and approved sources, ingredients may be contaminated when they arrive at the food establishment. It is important to know how to handle these ingredients safely and how to prepare food in such a manner that reduces the risk of contaminated food being served to your clients.

The Food and Drug Administration (FDA) is one of the federal agencies that is responsible for protecting our food supply. The Agency recognizes the importance of proper food handling in retail food establishments. It recommends that retail food managers be able to demonstrate knowledge in food safety.

Food safety in retail food establishments begins with managers who are knowledgeable about food hazards and who are committed to implementing proper food handling practices in their facility. It continues with properly trained food workers who understand the essentials of food safety and sanitation and who will not take short-cuts when it comes to food safety.

The authors of this textbook have been training retail food managers and employees for over 25 years. Many excellent resources are available for this type of training. However, the authors wanted to create a curriculum that was versatile enough to serve all segments of the retail food industry including restaurants, supermarkets, convenience stores, institutional facilities, and vending companies.

The essentials materials have been proven effective for teaching food safety and sanitation to many different audiences. The authors recommend the textbook and study guide for all of the following training activities:

- Short courses for retail food establishment managers to take in preparation for a national food protection manager certification examination,
- Food safety and sanitation courses in vocational and culinary arts programs,
- As an introductory course for Food Science Programs in colleges and universities,
- As a self-study program for retail food managers who are preparing to take a national food protection manager certification examination.

One of the most important tasks you face is to train and supervise food workers. Your knowledge is useless if you do not teach employees the correct way to handle food. Learn to recognize any break in standard operating procedure that might endanger food safety. Avoid the problems related to embarrassment, loss of reputation, and financial harm that accompany a foodborne disease outbreak.

Some state and local jurisdictions have passed legislation to require certification of one or more food managers in each establishment. Other jurisdictions are considering doing the same. Most certification programs require candidates to pass a written examination to measure knowledge of food safety and sanitation principles and practices. Some jurisdictions require completion of a food safety course before taking the exam.

Several different examinations are available for certification. In some cases, state and county regulatory agencies have developed their own examinations that apply specifically to their jurisdictions. There is, however, growing support for a nationally recognized examination and credential for food safety. If a national test is recognized by local agencies, a local examination might not be the better choice.

The Conference for Food Protection recognizes food protection manager examinations from the following provides:

▲ National Registry of Foods Safety Professionals of Professional Testing, Inc.

▲ Educational Foundation (EF) of the National Restaurant Association

▲ Experior (formerly National Assessment Institute, Block and Associates, and Insurance Testing Corporation)

Addresses for these groups are listed in Chapter 10.

When you want or need to become certified, check with your local regulatory agency to learn what they require. If you are pursuing a degree in hospitality management, culinary arts, or a related field ask your teacher or advisor for information and recommendations.

A card to request information or a test from the National Registry of Food safety Professionals is included with this book. The authors selected this examination because, like this text, it is based on the Food and Drug Administration's model *Food Code* which sets standards for all types of food establishments. The exam is also keyed to a new job task analysis derived from a survey sent to 2,500 food industry professionals.

The team of McSwane, Rue, and Linton wanted to produce a book to accompany the FDA's model *Food Code*. We trust that you will find this text accurate, comprehensive, and, most of all, useful.

David Z. McSwane, H.S.D., REHS, CFSP, is an Associate Professor of Public and Environmental Affairs at Indiana University. He has over 25 years of experience in food safety and sanitation working in state and local regulatory agencies and as a consultant to the food industry. Dr. McSwane has published numerous articles and presented papers on a variety of subjects related to food safety.

Dr. McSwane is a nationally recognized trainer in food safety and sanitation. He has taught courses at the university level and for regulatory agencies, food establishments, food industry trade associations, vocational schools, and environmental health associations throughout the United States. Dr. McSwane is a recipient of the Walter S. Mangold award. This is the highest honor bestowed by the National Environmental Health Association. He has been a correspondent for the *Food Protection Report* and is a member of the Environmental and Public Health Council at Underwriters Laboratories Inc. and the National Automatic Merchandising Association's Health Industry Council.

Nancy Roberts Rue, Ph.D., R.N., has a background in teaching in technical education. Her doctorate is in educational leadership and curriculum and instruction from the University of Florida. With thirty years of experience in higher education and evaluation at Indiana University and St. Petersburg Junior College, she is dedicated to the task of building educational materials that meet the needs of those who want to learn.

Dr. Rue wrote the *Handbook for Safe Food Service Management* while serving as director of the Certified Professional Food Manager program at National Assessment Institute. The *Handbook* was designed to give entry level food managers a single source for study and focused on need-to-know information. She is now an independent writer and consultant on training and development techniques.

Richard Linton, Ph. D., is an Associate Professor of Food Safety at Purdue University. His expertise is in the development and implementation of food safety and food quality programs, specifically in Hazard Analysis Critical Control Point (HACCP) systems. He is currently serving as chair of the HACCP Committee of the Conference for Food Protection. Dr. Linton spent twelve years working in retail food establishments. In recent years, he has educated food workers and managers from all segments of the industry throughout the nation and the world.

Dr. Linton has developed several different types of food safety training programs for retail food managers and workers. He also works closely with the retail food industry on research projects that help improve the quality and safety of the food they serve.

There are several reference books available in the field of food safety. However, most of these resources are directed toward a particular type of food establishment. The authors of *Essentials of Food Safety and Sanitation* believe there is need for a text that applies food safety principles to all food establishments, regardless of type. Chapter 4, *The Flow of Food*, is the focal point. It clearly defines the important strategies for handling food from receiving until it is placed in the hands of the consumer. Chapter 5 follows with a comprehensive description about how to apply the Hazard Analysis Critical Control Point (HACCP) system to the process.

The authors wish all readers continued success in their food safety and sanitation activities. Regardless of where you work, you must always remember—*foodborne illness is a preventable disease*. Follow the basic rules of food safety and head for a satisfying career in the food industry.

ACKNOWLEDGMENTS

The Authors wish to thank their families and colleagues who through their support and patience, helped us write the second edition of this book. John Wise continues to be a tremendous asset to this project. His drawings enable us to present information that is technical and sometimes tedious in a way that make it much less intimidating. Anna Graf Williams, Ph.D., and Karen J. Hall from Learnovation, camera-ready publishing and developmental editors, gave us the extra advice and encouragement that made the book come alive.

The Authors wish to thank Mary Jo Dolasinski for her contribution in writing the chapter on education and training. Arthur L. Banks and John Marcello were most gracious to assist in the review of selected chapters on very short notice. Ann Guentert was great assistance during the editing process, and Tom Campbell provided many of the photographs used in the text. Jeannine Smith provided valuable assistance with the computer-generated artwork.

1 FOOD SAFETY AND SANITATION MANAGEMENT

Foodborne Illness Can Be Hazardous to Your Health

Thirty-one people became ill with Salmonella after eating sandwiches prepared at a local fast-food establishment. A foodborne disease investigation of this outbreak showed that sandwiches topped with vegetable condiments, especially hamburgers, were most commonly associated with the illness.

The main consumers for plain hamburgers and cheeseburgers are children. However, there were no confirmed cases of Salmonella among children younger than 12 years old. This fact virtually eliminated buns and beef as the contaminated ingredients of the sandwiches. The fact that sandwiches with vegetable condiments were strongly associated with the illness suggests that either the vegetables were contaminated during preparation or as they were being placed on the sandwiches.

*A **foodborne disease outbreak** occurs when two or more people become ill after eating a common food. Salmonella bacteria cause several foodborne disease outbreaks each year.*

Learning Objectives

After reading this chapter, you should be able to:

▲ Recognize the importance of food safety and sanitation as the basis for preventing foodborne illness in retail food establishments

▲ State the problems caused by foodborne illness for both the individuals who become ill and the food establishment blamed for the incident

▲ Identify trends in menus and consumer use of food products prepared in food establishments

▲ Explain the role of government regulation (federal, state, and local) in retail food safety

▲ Recognize the types of food establishments used in this text and the influence of the *Food Code* on these operations

▲ Recognize the term Hazard Analysis Critical Control Point (HACCP) as a system for food protection

▲ Recognize the need for food protection manager certification.

Essential Terms

Clean	Foodborne Illness
Contamination	Microorganisms
	Germs
Food Code	Microbes
Food establishments	Sanitation
Foodborne disease outbreak	Sanitary

New Challenges Present New Opportunities

The food industry is one of America's largest enterprises. It employs about one-quarter of the nation's workforce and produces 20% of America's Gross Domestic Product (GDP). Billions of dollars worth of food are sold each year. Americans have made food a prominent part of their business and recreational activities.

The food industry is made up of businesses that produce, manufacture, transport, and distribute food for people in the United States and throughout the world. Food production involves the many activities that occur on farms, on ranches, in orchards, and in fishing operations. Food manufacturing takes the raw materials harvested by producers and converts them into forms suitable for distribution and sale. The retail distribution system consists of the many food operations that store, prepare, package, serve, vend, or otherwise provide food for human consumption. The term **food establishment** refers to facilities that are involved in food distribution. A list of food establishments is presented in Figure 1.1.

Figure 1.1—Types of Food Establishments

Food establishments, and the clients they serve, are very different. As you study problems related to food safety, keep this diversity in mind.

Surveys by the National Restaurant Association show that about 50 billion meals are eaten in restaurants and school and work cafeterias each year (The National Restaurant Association *Pocket Fact Book*). Large supermarkets carry an average of 30,000 items in their inventory.

The size and diversity of the food industry has made it virtually impossible for regulatory officials to continuously monitor all aspects of food safety. Therefore, food safety programs must involve governmental agencies and the food industry working closely together to ensure the safety of our food supply.

Food Safety—Why All the Fuss?

Everyone knows that the United States has one of the safest food supplies in the world. So why all the fuss? The answer is simple. Foodborne illness happens, and it adversely affects the health of millions of Americans every year. Foodborne illness is the sickness that some people experience when they eat contaminated food. It impairs performance and causes discomfort. Estimates of the number of cases of foodborne illnesses vary greatly. A recent report issued by the Centers for Disease Control and Prevention (CDC) estimates that foodborne diseases cause approximately 76 million illnesses, 325,000 hospitalizations, and 5,000 deaths in the United States each year (Mead et. al.).

Foodborne illness also has a major economic impact. It costs our society billions of dollars each year. These costs occur in the form of medical expenses, lost work, and reduced productivity by victims of the illness, legal fees, punitive damages, increased insurance premiums, lost business, and loss of reputation. Consumer confidence also goes down when a foodborne disease outbreak is reported. Food establishments that have been linked to a foodborne illness are likely to receive more frequent inspections by regulatory agencies. The establishment can also be closed if it is determined that is the best way to protect public health. A foodborne disease outbreak will have a negative impact on the owners, managers, and employees of the food establishment where it occurred and on the rest of the food industry as well.

Why Me?

You may be asking yourself, "What does all of this have to do with me?" The answer is PLENTY. Customer opinion surveys show that cleanliness and food quality are the top two reasons people use when choosing a place to eat and shop for food. Consumers expect their food to taste good and *not* make them sick. It is the responsibility of every food establishment owner, manager, and employee to prepare and serve safe and wholesome food and preserve their clients' confidence.

The Centers for Disease Control and Prevention (CDC) reports that in most foodborne illness outbreaks, the mishandling occurred within retail food establishments (restaurants, supermarkets, schools, churches, camps, institutions, and vending locations) where foods are prepared and served to the public. These foods may be eaten at the food establishment or taken out for consumption at another location.

The nature of food and the extent to which it is handled make the opportunity for contamination commonplace. Most cases of foodborne illness in retail food establishments are caused by foods that have been:

▲ Exposed to unsafe temperatures

▲ Handled by infected food workers who practice poor personal hygiene

▲ Exposed to harmful agents through cross contamination.

Each of these factors will be discussed in more detail in later chapters of this book.

As a food manager, you must understand that **foodborne illness can be prevented if the basic rules of food safety are routinely followed**. Figure 1.2 shows some of the key elements of food safety.

Prevention of foodborne illness must be a goal in every food establishment.

Figure 1.2—Elements of Safe Food

Changing Trends in Food Consumption and Choices

Today's menus reflect changing appetites among Americans. Ethnic foods are becoming quite popular, and their exotic ingredients often create new food safety challenges. Whether it's Thai food at a local restaurant or sushi sold at the supermarket, people have a wider variety of food to choose from in restaurants, grocery stores, and all other places where we obtain prepared food.

Due to changes in eating habits and more knowledge about food safety hazards, recommendations for safe food handling are changing. For example, we used to recommend cooking ground beef to an internal temperature of 140°F (60°C) to destroy dangerous bacteria that might be present. But that was before *E. coli* O157:H7 bacteria came on the scene. Now the recommended cooking temperature for ground beef is 155°F (68°C) or above. This higher temperature is needed to destroy the *E. coli* O157:H7 bacteria that may be present in the raw meat. Knowledge of basic food safety principles, as applied to different ingredients and methods of packaging, is critical to safe food management.

Consumers have less time to prepare food because more of them are working outside the home. As a result, they are buying more ready-to-eat food or products that require little handling. More than half of the food eaten by Americans is "fixed" in food processing plants, restaurants, supermarkets, delicatessens, cafeterias, institutions, and other sites outside the home. These foods are produced using a variety of processing, holding, and serving methods. All of these methods help protect foods from contamination.

The Problem: Foodborne Illness

Foodborne illness is a disease caused by the consumption of contaminated food. A **foodborne disease outbreak** is defined as an incident in which two or more people experience a similar illness after eating a common food.

Recent outbreaks of foodborne illness have been caused by:

▲ *E. coli* O157:H7 bacteria in lettuce, unpasteurized apple juice, and radish sprouts

▲ *Salmonella spp.* in alfalfa sprouts, ice cream, and dry cereal

▲ Hepatitis A virus in raw and lightly cooked oysters

▲ *Listeria monocytogenes* in hot dogs and luncheon meats.

The leading factors that contribute to foodborne illness are temperature abuse, poor personal hygiene practices, and cross contamination.

A growing number of people are especially susceptible to foodborne illness. These include:

▲ The very young

▲ The elderly

▲ Pregnant or lactating women

▲ People with impaired immune systems due to cancer, AIDS, diabetes, or medications that suppress response to infection.

Foodborne illness can cause severe reactions, even death, to those individuals in these high risk categories. The availability of a safe food supply is critical to these people.

Outbreaks of foodborne illness receive widespread media attention. Newspapers, television, and radio have become sensitive to issues surrounding food safety. As a result, public awareness about the problem, and how it occurs, has increased dramatically in the last few years.

Contamination

Whether a food item is prepared "from scratch" or arrives in a ready-to-eat form, there are opportunities for it to become contaminated before it is consumed. **Contamination** is the presence of substances or conditions in the food that can be harmful to humans. Figure 1.3 shows the agents that cause most foodborne illnesses. **Bacteria poses the greatest food safety challenge for all retail food establishments.**

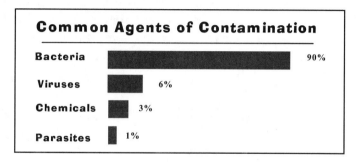

Figure 1.3—Types of Contamination

(Source: Surveillance for Foodborne Disease Outbreaks—U.S., 1988-1992)

Foods can become contaminated at a variety of points as the food flows from the farm to the table (Figure 1.4). Raw foods can be contaminated at the farm, ranch, or on board a ship. Contamination can also occur as foods are handled during processing and distribution. Measures to prevent and control contamination must begin when food is harvested and continue until the food is consumed.

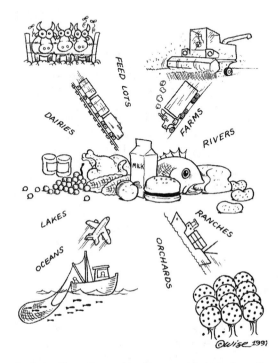

Figure 1.4—Foods Can Become Contaminated at Several Points Between the Farm and the Table.

Figure 1.5 shows some of the more common sources of contamination such as soil, water, air, plants, animals, and humans. Contaminants present an "invisible challenge" because you can't see them with the naked eye. Many types of food contamination can cause illness without changing the appearance, odor, or taste of food.

Germs can be transferred from one food item to another by **cross contamination**. This typically happens when microbes from a raw food are transferred to a cooked or ready-to-eat food via contaminated hands, equipment, or utensils.

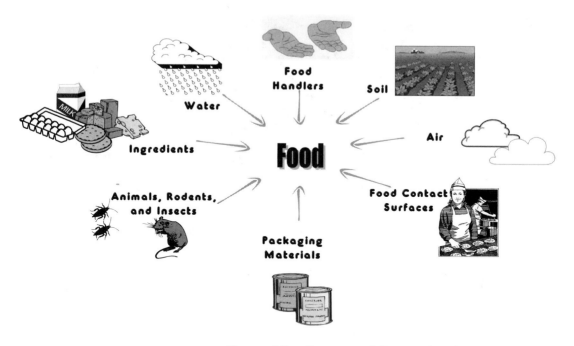

Figure 1.5—Sources of Contamination

Microorganisms (Germs or Microbes)

Microorganisms (also called **germs** or **microbes**) are the most common types of food contamination. Microorganisms include bacteria, viruses, parasites, and fungi that are so small they can only be seen with the aid of a microscope. Microbes are everywhere around us—in soil, water, air, and in and on plants and animals (including humans). The microorganisms that cause most foodborne illnesses are bacteria. These organisms get their energy from the food in which they live and reproduce—often the same foods we eat.

Most microorganisms are harmless. However, some microbes can cause problems when they get into food. The microbes that must be controlled in a food establishment are the ones that cause foodborne illness and food spoilage. Figure 1.6 shows three of the most common types of bacteria which cause foodborne illnesses. It is important to remember that germs that cause foodborne illness usually do not alter the taste, odor, and appearance of the food they are in.

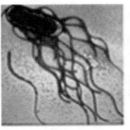
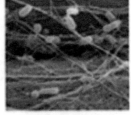
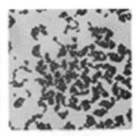

a. *Salmonella* bacteria   b. *E.coli* bacteria   c. *Staphylococcus aureus* bacteria

Figure 1.6— Common Causes of Foodborne Illness
(Source: *www.fightbac.org*)

The Food Flow

Food flow consists of food products, and the ingredients used to make them, as they "flow" through a food establishment. The food flow begins with the purchase of safe and wholesome ingredients from approved sources. Once the food is delivered, it then flows through receiving into storage. The final stages in the food flow are preparation and service. Preparation and service include all the activities that occur between storage and consumption of the food by your customers.

Preparation steps frequently involve thawing, cooking, cooling, reheating, hot-holding, and cold-holding. Improper food preparation and service can lead to foodborne illness. For potentially hazardous foods (those that support bacterial growth), time and temperature must be monitored and controlled. Preparation of foods not requiring cooking (ready-to-eat foods) can also involve a lot of contact with your hands or with food contact surfaces. It is always important for you, the food worker, to practice good personal hygiene and use proper handwashing techniques. However, during preparation and service it is crucial!

Since preparation and service are typically the last steps prior to consumption of the food, efficient monitoring and control of safe food practices are also critical. As products flow through a food establishment they are handled many times and in many different ways. For all food establishments, controlling food safety and quality is of utmost importance. You must learn to recognize foods that may have been contaminated prior to delivery and understand what is required to keep these products safe until you serve them to your customers.

A New Approach to an Old Problem

Food industry professionals and regulatory officials agree that better ways to protect people from foodborne illness must be found. The *Food Code* recommends using the Hazard Analysis Critical Control Point (HACCP) system as a means of ensuring food safety. This innovative system was developed by the Pillsbury Company in the 1960s for the NASA space program. Officials in the space program realized that a foodborne illness in space could present a life threatening situation. Therefore, it was critical that all food used by the astronauts be as near 100% safe as possible.

The HACCP system follows the flow of food through the food establishment and identifies each step in the process where contamination might cause the food to become unsafe. When a problem step is identified, action is taken to make the product safe or, if that is not feasible, the food is discarded. Individual food facilities can control the flow of food by setting up an in-house HACCP program. The HACCP program should be designed to accommodate the types of products served and the production equipment and processes used in the establishment. Records used in conjunction with the HACCP system are kept on file at the food establishment. These records will be periodically reviewed by regulatory inspectors to ensure that the system is working properly.

HACCP systems for monitoring food safety have been used in food manufacturing plants for many years. Now this system is being recommended for use in food establishments as well. A more detailed discussion of HACCP system is presented in Chapter 5.

Facility Planning and Design

A well-planned facility with a suitable layout is essential for the smooth operation of any food establishment. Layout, design, and facilities planning directly influence:

▲ Worker safety and productivity

▲ Labor and energy costs

▲ Customer satisfaction.

Some food establishments are located in buildings that were designed and constructed specifically for a food operation. Others are located in buildings that have been converted to accommodate food preparation and service activities. Either way, the better your facility is planned, the easier it will be to achieve your food safety goals and earn a profit.

Keeping It Clean and Sanitary

Customer opinion surveys show that cleanliness is a top consideration when choosing a place to eat or shop for food. Customers' impressions about the cleanliness of an operation are frequently influenced by what they see inside and outside the establishment. Inside and outside, the facility must be clean and free of litter. Customer satisfaction is highest in food establishments that are clean and bright and where quality food products are safely handled and displayed. By increasing customer satisfaction, you will develop a devoted clientele who will regularly patronize your establishment.

It is the responsibility of every person working in the food industry to keep things **clean** and **sanitary**. Effective cleaning of equipment reduces the chances of food contamination during preparation, storage and service (see Figure 1.7). **Cleaning** is concerned with the removal of visible soil from the surfaces of equipment and utensils. **Sanitary** means healthful or hygienic. It involves reducing the number of disease-causing microorganisms on the surface of equipment and utensils to acceptable public health levels. Something that is sanitary poses little or no risk to human health.

Good sanitation minimizes attraction of pests, increases life of equipment, improves employee morale and efficiency, and is important from other aesthetic considerations.

Poor performance by the cleanup crew may result in dirty utensils and equipment which could harm customers and hinder business. Sanitation practices must be employed at all times to protect the health and well-being of customers.

CLEAN

SANITARY

Figure 1.7—Clean Looks Good—Sanitary Makes Sure It's Safe

Accident Prevention and Crisis Management

An accident prevention program is a must in every food establishment. The cost of accidents may mean the difference between profit and loss. Beyond financial responsibilities, the loss of a worker's skills can disrupt operations and cause additional stress on other employees. Ensuring a safe environment for employees and customers requires continuous monitoring, but the rewards make it worth the effort.

Human error always will be a factor in food establishments, but training and proper equipment can help employees avoid accidents. Pan handles sticking over the edge of the stove, poorly stacked boxes in hallways or store rooms, spilled food, and many other situations can lead to serious injuries. For example, a fall on a wet floor can cause expenses running into the thousands of dollars for lost time, worker's compensation, and medical care.

In food establishments, another kind of crisis occurs when water supplies, electricity, or sewer systems are disrupted. Food codes contain specific instructions on how to operate when basic services are lost. Public health departments can help when you are not sure how to proceed.

Storms, tornados, hurricanes, or floods cause conditions that require changes to maintain food service. A good disaster plan is invaluable in times of need. Managers and supervisors are expected to know how to handle these and other emergencies, when they occur.

Education and Training Are Key to Food Safety

The prevention of foodborne illness begins with the knowledge of where contaminants come from, how they get into food, and what can be done to control or eliminate them. You are not expected to become a food scientist, but a working knowledge of what causes foodborne illness is valuable information. You must know the correct way to manage food safety and sanitation, and your employees must know correct food handling procedures.

The importance of teaching employees about food safety is heightened because of the global nature of our food supply. Control of factors during growth, harvest, and shipping is not always possible when food is produced in so many different parts of the world. Also, an error in time/temperature management, cross contamination, or personal health and hygiene of food workers can cost a life among our vulnerable citizens. As you have learned, storage, preparation, holding, and service procedures are critical in the prevention of foodborne illness. Employees do not typically come to the job knowing this information. They have to be trained.

Managing the safety of food involves controlling the supply, maintaining sanitary facilities, and training employees to know how to work with food safely. Knowledge, skills, and abilities must focus on food safety practices that will protect the public from foodborne illness. Food safety programs require food managers and food workers to have the knowledge and skills needed to keep food safe. It is not enough that the manager knows food safety—employees too must be taught how to perform correctly.

The Role of Government in Food Safety

The purpose of government regulation in food safety is to oversee the food-producing system and protect food that is intended for human consumption. Governmental agencies enforce laws and rules to protect food against adulteration and contamination. Regulatory personnel monitor both the process and the product to ensure the safety of the food we eat.

Federal, state, and local regulators inspect food products and production facilities as a way of detecting problems. However, there are not enough inspectors and other resources to adequately administer the program. The enormous size of the food industry makes it virtually impossible for regulatory inspectors to effectively monitor all aspects of food safety. Therefore, food safety programs involve governmental agencies and the food industry working together to ensure the safety of our food supply.

The Food Code

In 1993, the U.S. Food and Drug Administration (FDA) published the *Food Code* which replaced three earlier model codes—*The Food Service Sanitation Manual Including a Model Food Service Sanitation Ordinance of 1976*, *The Vending of Food and Beverages Including a Model Sanitation Ordinance of 1978*, and *The Retail Food Store Sanitation Code of 1982*—with one code. The FDA has made a commitment to revise the *Food Code* every two years, with the assistance of experts from state and local government, industry, professional associations, and colleges and universities. It was determined that a biannual revision was necessary to keep pace with changes that are occurring in food safety and the food industry. In this text, the *Food Code* will be used as the point of reference.

The *Food Code* is not a law. Rather, it is a set of recommendations intended to be used as a model by state and local jurisdictions when formulating their own rules and regulations. You will need to obtain a copy of the health rules and regulations that apply to the food establishments in your jurisdiction.

The Role of the Food Industry in Food Safety

The food industry has accepted greater responsibility for overseeing the safety of its own processes and products. Consumers expect and deserve food that is safe to eat. If a food establishment is involved in a foodborne disease outbreak, consumers may retaliate by taking their business elsewhere or by seeking legal action. Financial loss and damaged reputation are some of the outcomes of a foodborne disease outbreak that can cause serious harm to the food establishment found responsible for the problem. One means of preventing the harmful effects of a foodborne disease outbreak is to start a food safety assurance program in the food establishment. This helps ensure that proper safeguards are used during food production and service. The ability to prove that a food safety system was in place at the time a foodborne disease outbreak occurred is very important. It has been deemed an acceptable defense in court cases where punitive damages have been sought by victims of foodborne illness.

Food Protection Manager Certification

Several organizations including the Conference for Food Protection, the U.S. Food and Drug Administration, and the trade associations for food retailers have recommended food manager certification as a means of ensuring that responsible individuals in charge of food establishments are knowledgeable about food safety. A **certified food protection manager** is a person who:

▲ Is responsible for identifying hazards in the day-to-day operation of a food establishment that prepares, packages, serves, vends or otherwise provides food for human consumption

▲ Develops or implements specific policies, procedures, or standards aimed at preventing foodborne illness

▲ Coordinates training, supervises or directs food preparation activities, and takes corrective action as needed to protect the health of the consumer

▲ Conducts in-house self-inspection of daily operations on a periodic basis to see that policies and procedures concerning food safety are being followed.

A certified food protection manager will have to be able to demonstrate knowledge and skills in food protection management including the following areas:

▼ Identifying of foodborne illness

▼ Describing the relationship between time and temperature and the growth of microorganisms that cause foodborne illness

▼ Describing the relationship between personal hygiene and food safety

▼ Describing methods for preventing food contamination from purchasing and receiving

▼ Recognizing problems and potential solutions associated with facility, equipment, and layout in a food establishment

▼ Recognizing problems and solutions associated with temperature control, preventing cross-contamination, housekeeping and maintenance.

The Conference for Food Protection has adopted a complete list of knowledge areas that is presented in Appendix E.

Some state and municipal agencies already require certification of food managers, and others are considering it. Most certification programs require managers to pass a written food safety examination. These examinations are designed to link food safety theory to practice. By passing a certification examination, the food manager is able to demonstrate proficiency in food protection management. The following organizations/agencies offer food safety certification exams:

The National Registry of Food Safety Professionals
1200 Hillcrest Street, Suite 300
Orlando, FL 32803-4717
1-800-330-3776
Fax: 407-894-1646
pti@proftesting.com

Experior (formerly NAI Block and Insurance Testing Corp.)
600 Cleveland Street, Suite 900
Clearwater, FL 33755
727-449-8525
Fax: 727-461-2746

Educational Foundation of the National Restaurant Assn.
250 S. Wacker Drive, Suite 1400
Chicago, IL 60606-5834
1-800-765-2122
Fax: 312-715-1217

Some managers will seek training before they take a certification examination. Some jurisdictions require training before testing, whereas others require only the test. Training can involve participating in a formal course, or it could involve using a non-traditional approach such as home study, correspondence, distance learning, and interactive computer programming.

This book has been written as a teaching tool to be used in conjunction with traditional and non-traditional food manager certification programs. Like the manager certification examinations, it links theory to practice. The authors of this book recognize that knowing about food safety is not enough. You must also be able to apply the principles and practices that promote the safety of food preparation and service.

Summary

In the case study presented at the beginning of this chapter, you read about a foodborne disease outbreak caused by contaminated sandwich toppings. In this case, a food worker was infected with *Salmonella* bacteria. As is sometimes the case, the worker did not show any of the symptoms that are common to the illness.

The bacteria was spread by fecal-oral contamination. In other words, by not washing his hands properly after using the toilet, the worker transferred *Salmonella* bacteria from his hands to the vegetable condiments that he placed on the sandwiches. This case illustrates the importance of good personal hygiene, especially handwashing, as a means of preventing the spread of germs that cause foodborne illness.

Everyone who works in a food establishment must understand that foodborne illness is *preventable*. It is the duty of every food establishment operator, manager, and employee to serve safe and wholesome food to their customers. Failure to do so can have serious financial impact on your establishment, and may cost you your job.

You can protect the health and safety of your customers by developing and implementing effective food safety and sanitation practices within your establishment. In the following chapters, you will learn more about how food is contaminated and what actions are needed to prevent, control and eliminate the agents that frequently cause foodborne illness and spoilage.

Case Study 1.1

Fourteen people became ill after eating pie that had been highly contaminated with *Salmonella enteritidis*. Several of the victims were hospitalized, and a man in his forties, who was otherwise in good health, died as a result of the foodborne illness.

In this outbreak, cream, custard, and meringue pies were made using ingredients from shell eggs. The pies were baked in a restaurant bakery and were stored for two and a half hours in a walk-in cooler before being transported in the trunk of a car to a private company outing. The pies were consumed three to six hours later. Leftover pie was consumed later that evening and the next day after having been kept unrefrigerated for as long as 21 hours.

▲ What conditions may have promoted bacterial growth?

Answers to the case study can be found in Appendix A.

Discussion Questions

1. Discuss the role of government in food safety programs.

2. Define the terms *sanitation* and *contamination.*

3. List some examples of *food establishments.*

4. How does news media coverage influence food safety issues?

5. What is a *foodborne disease outbreak?*

6. What impact do current menu item trends have on food safety?

7. How does food safety affect the young, elderly, pregnant women, and those with immune system problems?

8. Discuss how certification of food managers could affect food safety?

9. Identify some reasons to implement food safety programs in food establishments.

10. Discuss some reason(s) you are studying food safety and sanitation.

Quiz 1.1 (Multiple Choice)

1. Which of the following groups is **not** especially susceptible to foodborne illness?

    a. The very young.

    b. Young adults.

    c. The elderly.

    d. Pregnant or lactating women.

2. The cost of foodborne illness can occur in the form of:

    a. Medical expenses.

    b. Loss of sales.

    c. Legal fees and fines.

    d. All of the above.

3. CDC reports show that in most foodborne illness outbreaks, mishandling of the suspect food occurred within which of the following stages?

    a. Transportation.

    b. Retail.

    c. Food manufacturing.

    d. Food production (farms, ranches, etc.).

4. Popular food trends and a larger variety of food items from around the world along with new methods of preserving or packaging food:

    a. Pose new challenges for food safety practices.

    b. Have no impact on food safety issues.

    c. Decrease the need for new food processing equipment.

    d. Are passing fads and require little adjustment in food handling practices.

5. How does the *Food Code* affect individual states and jurisdictions?

    a. The *Code* is a Federal law that must be enforced by state agencies.

    b. The *Code* regulates food manufacturing facilities (processors) in state jurisdictions.

    c. It provides a model for new laws and rules in state and local jurisdictions.

    d. It validates current practices.

Quiz 1.2 (True/False)

Answer questions 1-5 with either True or False.

1. The Hazard Analysis Critical Control Point System is only used to monitor food processing in manufacturing plants.

2. Certification of food protection managers and workers refers to screening done for health problems.

3. Regulatory agencies are assigned the task of monitoring the production and processing of food from harvest to the consumer.

4. The term Food Establishment includes any site where food is processed, prepared, sold, or served.

5. Bacteria cause most foodborne illnesses.

Answers to the multiple choice and true/false questions are provided in Appendix A.

References/Suggested Readings

Banwart, G.J. (1989). *Basic Food Microbiology*. Van Nostrand and Rheinhold. New York, NY.

Centers for Disease Control and Prevention (1996). *Surveillance for Foodborne Disease Outbreaks—United States, 1988-1992*. U.S. Department of Health and Human Services. Atlanta, GA.

Food and Drug Administration (1999). *1999 Food Code*. United States Public Health Service. Washington, DC.

Mead, Paul S., Laurence Slutsker, Vance Dietz, Linda F. McCaig, Joseph S. Bresee, Craig Shapiro, Patricia M. Griffin, and Robert V. Tauxe (1999). "Food-Related Illness and Death in the United States". *Emerging Infectious Diseases*, Vo. 5 No. 5, September-October, 1999. Centers for Disease Control and Prevention, Atlanta, GA.

National Restaurant Association (1997). *Pocket Fact Book*. Washington, DC.

Ronsivalli, L.J., and E.R. Vieira (1992). *Elementary Food Science*. Champan and Hall. New York, NY.

Potter, N., and J. Hotchkiss (1995) *Food Science* (3rd ed.). Chapman and Hall. New York, NY.

Rue, N., and National Assessment Institute (1994). *Handbook for Safe Food Service Management*. Prentice Hall. Upper Saddle River, NJ.

Spears, M.C. (1995). *Foodservice Organization: A Managerial Approach* (3rd edition). Prentice Hall. Upper Saddle River, NJ.

Suggested Web Sites

Gateway to Government Food Safety Information
www.foodsafety.gov

United State Department of Agriculture (USDA)
www.usda.gov

Centers for Disease Control and Prevention (CDC)
www.cdc.gov

United States Food and Drug Administration (FDA)
www.fda.gov

USDA/FDA Food and Nutrition Information Center
www.nal.usda.gov/fnic/

United States Environmental Protection Agency (EPA)
www.epa.gov

Partnership for Food Safety Education
www.fightbac.org

The Food Marketing Institute
www.fmi.org

National Restaurant Association
www.restaurant.org

National Registry of Food Safety Professionals
www.nrfsp.com

2 HAZARDS TO FOOD SAFETY

Pesticide Mistaken as Black Pepper - Makes Employees Very Sick

In the summer of 1998, 20 employees attended a company lunch prepared from homemade foods. Soon after the meal, 14 people experienced a variety of symptoms including stomach cramps, nausea, diarrhea, dizziness, sweating, and muscle twitching. Ten victims visited a hospital emergency department, and two were hospitalized.

The foods served at the lunch included roasted pork, boiled rice, cabbage salad, biscuits and soft drinks. Sixteen of the 20 employees who attended the luncheon ate the cabbage salad; of these 14 became ill. The four people who did not eat the cabbage salad did not develop symptoms.

The employee who prepared the cabbage salad reported mixing two 1-lb bags of precut, prepackaged cabbage in a bowl with vinegar and ground pepper. The pepper came from a can labeled "black pepper" that he had found 6 weeks before the lunch in the truck of deceased relative. The employee had not used the pepper prior to fixing the cabbage salad.

The contents of the black pepper container and the cabbage salad were tested for toxic chemicals. Both were found to contain aldicard—one of the most potent pesticides in America. The deceased owner of the pepper can had been a crawfish farmer. Foodborne disease investigators believe the farmer had used aldicard on bait to prevent destruction of his crawfish nets, ponds, and levees by wild dogs and raccoons.

Learning Objectives

After reading this chapter, you should be able to:

▲ Identify and give examples of each of the three main types of foodborne hazards

▲ Discuss how the three classes of foodborne illness cause disease

▲ List the factors that promote bacterial growth in foods

▲ Identify the food temperature danger zone

▲ Explain how temperatures in the danger zone affect bacterial growth

▲ Identify the major types of potentially hazardous foods and the characteristics that are common to this group of foods.

Essential Terms

At risk	FATTOM
Acidic	Food allergen
Alkaline	Infection
Bacteria Aerobic	Ingestion
Anaerobic Binary fission	Inherent
Facultative anaerobic Inherent	Intoxication
Mesophilic Psychrophilic	Onset time
Thermophilic Parasite	pH
Pathogenic Spore	Physical hazards
Vegetative state	Potable water
Biological hazards	Potentially hazardous foods
Chemical hazards	Ready-to-eat foods

Spoilage

Spore

Temperature abused

Temperature danger zone

Toxin

Toxin-mediated infection

Vegetative cells

Virus

Water activity (A_w)

Types of viruses:
  Hepatitis A
  Norwalk virus group

Types of bacteria:
Bacillus cereus
Campylobacter jejuni
Clostridium perfringens
Clostridium botulinum
Escherichia coli
Listeria monocytogenes
Salmonella spp.
Shigella spp.
Staphylococcus aureus
Vibrio spp.

Types of parasites:
  Anisakis spp.
  Cryptosporidium parvum
  Cyclospora cayetanensis
  Giardia lamblia
  Toxoplasma gondii
  Trichinella spiralis

Foodborne Illness

Have you ever had a foodborne illness? Many of us have, but it did not last long enough to be diagnosed. We think we have the "stomach flu" or a "24 hour bug" of some kind.

Did you become nauseated, feel abdominal pain and cramping, have diarrhea, and even start vomiting? It could be foodborne illness. When this happens, do you think it is from eating or drinking too much? If you are far from home, do you blame the "strange water"?

What you experience is often a foodborne illness (Figure 2.1). The symptoms are very similar to those associated with the flu. The type of microbe, how much contamination is in the food, and the general condition of the affected person determine the severity of the symptoms.

Figure 2.I—Ingredients for Foodborne Illness

General symptoms of foodborne illness usually include one or more of the following:

▲ Headache   ▲ Abdominal pain

▲ Nausea   ▲ Diarrhea

▲ Vomiting   ▲ Fatigue

▲ Dehydration   ▲ Fever

Foodborne illness is generally classified as a foodborne infection, intoxication, or toxin-mediated infection (Figure 2.2). Your awareness of how different microbes cause foodborne illness will help you understand how they contaminate food.

Infection	Caused by eating food that contains living disease-causing microorganisms.
Intoxication	Caused by eating food that contains a harmful chemical or toxin produced by bacteria or other source.
Toxin-Mediated Infection	Caused by eating a food that contains harmful microorganisms that will produce a toxin once inside the human body.

Figure 2.2—Classification of Foodborne Illness

When a living, disease-causing microorganism is eaten along with a food, it can cause a **foodborne infection**. After ingestion, the organism burrows into the lining of the victim's digestive tract and begins to grow in number. This can lead to the common symptoms of foodborne illness like diarrhea. Sometimes, the microbes may spread to other parts of the body through the blood stream. Bacteria, viruses, and parasites are examples of microorganisms that can cause infection. A common type of foodborne infection is *Salmonellosis.* This disease is caused by *Salmonella* bacteria that are frequently found in poultry and eggs.

An **intoxication** is caused when a living organism multiplies in or on a food and produces a chemical waste or toxin. If the food containing the toxin is eaten, the toxin causes an illness. Typically called food poisoning, common examples of food intoxications are *Clostridium botulinum* and *Staphylococcus aureus*. An intoxication may also occur when an individual consumes food that contains man-made chemicals such as cleaning agents or pesticides.

A **toxin-mediated infection** is caused when a living organism is consumed with food (as in the case of an infection). Once the organism is inside the human body it produces a toxin that causes the illness. A toxin-mediated infection is different from an intoxication because the toxin is produced inside the human body. An example of an organism that causes this type of illness is *Clostridium perfringens.*

Foodborne illnesses have different onset times. The **onset time** is the number of hours between the time a person eats contaminated food and when they first show symptoms of the disease. This is usually expressed in a range of hours since individual onsets vary depending on factors such as age, health status, body weight, and the amount of contaminant they ingested with the food. However, for some foodborne diseases, such as trichinosis and Hepatitis A, the onset time can be several days.

Under the right circumstances, anyone can become ill from eating contaminated foods. In most cases, healthy adults will have flu-like symptoms and recover in a few days.

However, the risks and dangers associated with foodborne illness are much greater for certain members of the population (see Figure 2.3). Those of high risk populations include infants and young children, the elderly, pregnant women, those with suppressed immune systems as occurs with acquired immune deficiency syndrome (AIDS), cancer, diabetes, and people taking certain medications. For these individuals, the symptoms and duration of foodborne illness can be much more severe—even life threatening.

Figure 2.3—High Risk Groups

Foodborne Hazards

A **foodborne hazard** refers to a biological, chemical, or physical hazard that can cause illness or injury when consumed along with the food (see Figure 2.4).

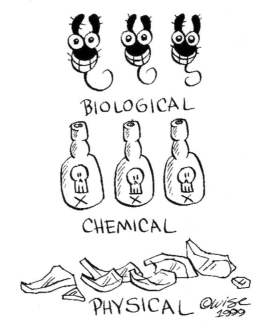

Figure 2.4—Foodborne Hazards

Biological hazards include bacteria, viruses, and parasites. These organisms are very small and can only be seen with the aid of a microscope. Biological hazards are commonly associated with humans and with raw products entering the food establishment. **Bacteria** are single-celled microorganisms that require food, moisture, and warmth to multiply. Bacteria can cause foodborne infections, intoxications, and toxin-mediated infections. Many of these organisms occur naturally in the environment where foods are grown. Most are destroyed by adequate cooking, and numbers are kept to a minimum by proper cooling during product distribution and storage.

Biological hazards are by far the most important foodborne hazard in any type of food establishment. They cause most foodborne illnesses and are the primary target of a food safety program.

Chemical hazards are toxic substances that may occur naturally or may be added during the processing of food. Examples of chemical contaminants include agricultural chemicals (i.e., pesticides, fertilizers, antibiotics), cleaning compounds, heavy metals (lead and mercury), food additives, and food allergens. Harmful chemicals at very high levels have been associated with severe poisonings and allergic reactions. Chemicals and other non-food items should never be placed near food items.

Physical hazards are hard or soft foreign objects in food that can cause illness and injury. They include items such as fragments of glass, metal, unfrilled toothpicks, jewelry, adhesive bandages, and human hair. These hazards result from accidental contamination and poor food handling practices that can occur at many points in the food chain from the source to the consumer.

Bacteria

Bacteria are the most important biological foodborne hazard for any food establishment. Bacteria are responsible for more cases of foodborne illness than any other hazard.

All bacteria exist in a vegetative state. **Vegetative cells** grow, reproduce, and produce wastes just like other living organisms.

Some bacteria have the ability to form spores. **Spores** help bacteria survive when their environment is too hot, cold, dry, acidic, or when there is not enough food. Spores are not able to grow or reproduce (see Figure 2.5).

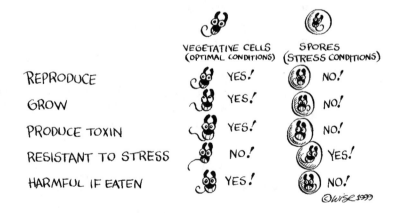

	VEGETATIVE CELLS (OPTIMAL CONDITIONS)	SPORES (STRESS CONDITIONS)
REPRODUCE	YES!	NO!
GROW	YES!	NO!
PRODUCE TOXIN	YES!	NO!
RESISTANT TO STRESS	NO!	YES!
HARMFUL IF EATEN	YES!	NO!

©WISE 1999

Figure 2.5—The Vegetative and Spore States of Bacteria Cells

However, when conditions become suitable for growth, the spore will germinate much like a seed. The bacterial spore can then return to the vegetative state and begin to grow again. Bacteria can survive for many months as spores. Also, it is much harder to destroy bacteria when it is in a spore form (Figure 2.6).

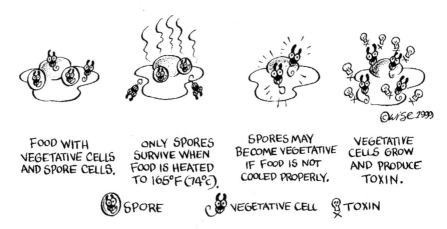

FOOD WITH VEGETATIVE CELLS AND SPORE CELLS.

ONLY SPORES SURVIVE WHEN FOOD IS HEATED TO 165°F (74°C).

SPORES MAY BECOME VEGETATIVE IF FOOD IS NOT COOLED PROPERLY.

VEGETATIVE CELLS GROW AND PRODUCE TOXIN.

SPORE   VEGETATIVE CELL   TOXIN

Figure 2.6—Spores and Toxins Pose a Special Challenge

Spoilage and Disease-Causing Bacteria

Bacteria are classified as either spoilage or pathogenic (disease-causing) microorganisms.

Spoilage bacteria degrade (break down) foods so that they look, taste, and smell bad. It reduces the quality of food to unacceptable levels. When this happens, the food will have to be thrown away.

Pathogenic bacteria are disease-causing microorganisms that can make people ill if they or their toxins are consumed with food.

Both spoilage and pathogenic bacteria must be controlled in food establishments. The steps you take to keep foods safe from disease-causing bacteria will also improve the culinary quality and shelf life of the food.

Bacterial Growth

Bacteria reproduce when one bacterial cell divides to form two new cells. This process is called **binary fission**. The reproduction of bacteria and an increase in the number of organisms is referred to as bacterial growth (Fig.2.7).

Figure 2.7—Division of Bacterial Cells

What Bacteria Needs in Order to Multiply

Bacteria need six conditions in order to multiply. They need a source of food, a mildly acid environment (pH of 4.6 - 7.0), a temperature between 41°F and 140°F (5° and 60°C), time, different oxygen requiring environments, and enough moisture. These requirements can be remembered using the acronym F-A-T-T-O-M (Figure 2.8). Since many foods naturally contain microorganisms, we need to be sure to control these six conditions as much as possible to prevent bacteria from multiplying.

Bacterial growth follows a regular pattern that consists of four phases (Figure 2.9). The first phase is the *lag phase* in which the bacteria exhibit little or no growth. The bacteria adjust to their surroundings during the lag phase. The lag phase lasts only a few hours at room temperature. However, the duration of this phase can be increased by keeping foods at 41°F (5°C) or below.

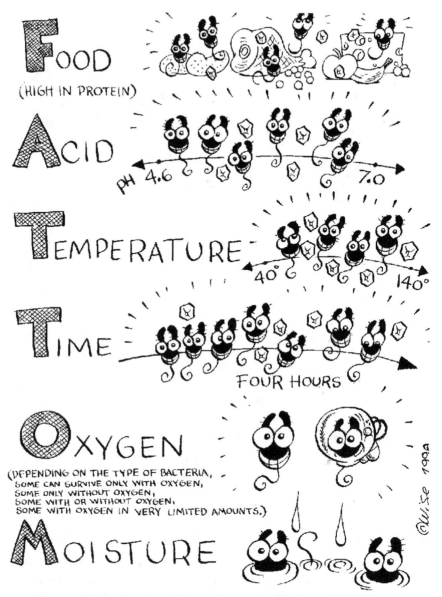

FOOD (HIGH IN PROTEIN)

ACID pH 4.6 — 7.0

TEMPERATURE 40° — 140°

TIME FOUR HOURS

OXYGEN (DEPENDING ON THE TYPE OF BACTERIA, SOME CAN SURVIVE ONLY WITH OXYGEN, SOME ONLY WITHOUT OXYGEN, SOME WITH OR WITHOUT OXYGEN, SOME WITH OXYGEN IN VERY LIMITED AMOUNTS.)

MOISTURE

Figure 2.8—Six Conditions Bacteria Need to Multiply

The second phase of bacterial growth is the *log phase*. Bacterial growth is very rapid during the log phase with bacteria doubling in numbers every few minutes. Keeping bacteria from reaching the log phase of growth is critical for food safety.

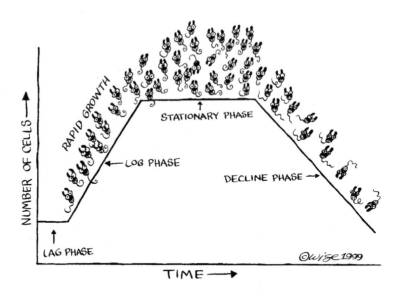

Figure 2.9—Bacterial Growth Curve

The third phase of bacterial growth is the *stationary phase*. The number of new bacteria being produced equals the number of organisms that are dying off during this phase. The bacteria have used up much of the space, nutrients, and moisture in the food by this phase of the growth curve.

The final phase in the growth curve is the *decline phase*. Here bacteria die off rapidly because they lack nutrients and are poisoned by their own toxic wastes.

Source of Food

A suitable food supply is the most important condition needed for bacterial growth. Most bacteria prefer foods that are high in protein like meats, poultry, seafood, dairy products, and cooked rice, beans, and potatoes (Figure 2.10).

Acidity

The **pH** symbol is used to designate the acidity or alkalinity of a food. You measure pH on a scale that ranges from 0 to 14 (Figure 2.11).

Figure 2.10—Microbes Eat the Same Food We Do

Most foods are acidic and have a pH less than 7.0. Very acid foods (pH of 4.5 or below), like lemons, limes, and tomatoes, will not normally support the growth of disease-causing bacteria. Pickling fruits and vegetables preserves the food by adding acids such as vinegar. This lowers the pH of the food in order to slow down the rate of bacterial growth.

A pH above 7.0 indicates the food is **"alkaline."** Examples of alkaline foods are olives, egg whites, or soda crackers.

Most bacteria prefer a neutral environment (pH of 7.0) but are capable of growing in foods that have a pH in the range of 4.6 to 9.0. Since most foods have a pH of less than 7.0, we have identified the range where harmful bacteria grow from 4.6 to 7.0. Many foods offered for sale in food establishments have a pH in this range.

Disease-causing bacteria grow best when the food it lives on has a pH of 4.6 to 7.0. Milk, meat, and fish are in this range.

Temperature

All bacteria do not have the same temperature requirements for growth. **Psychrophilic** bacteria grow within a temperature range of

32° to 70°F (0° to 21°C). These microorganisms are especially troublesome because they are capable of multiplying at both refrigerated and room temperatures. Most psychrophilic bacteria are spoilage organisms, but some can cause disease.

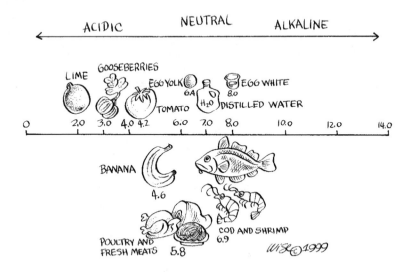

Figure 2.11—Acidity Range of Common Foods Prepared in Food Establishments

Mesophilic (middle range) bacteria grow at temperatures between 70° and 110°F (21° and 43°C), with most rapid growth at human body temperature (98.6°F, 37°C).

Bacteria that grow best at temperatures above 110°F (43°C) are called **thermophilic** organisms. All thermophilic bacteria are spoilage organisms.

Time and temperature are the most critical factors affecting the growth of bacteria in foods. Most disease-causing bacteria can grow within a temperature range of 41° to 140°F (5° to 60°C). This is commonly referred to as the food **"Temperature Danger Zone"** (Figure 2.12). Some disease-causing bacteria, such as *Listeria monocytogenes*, can grow at temperatures below 41°F (5°C), but the rate of growth is very slow.

Careful monitoring of time and temperature is the most effective way a food establishment manager has to control the growth of disease-causing and spoilage bacteria. A saying related to temperature control in the food industry is "Keep it hot, keep it cold, or don't keep it!" This means all cold foods must be stored at 41°F (5°C) or below and all hot foods held at 140°F (60°C) or above. **Temperature abuse** is the term applied to foods that have not been heated to a safe temperature or kept at the proper temperature. This could result in a foodborne illness.

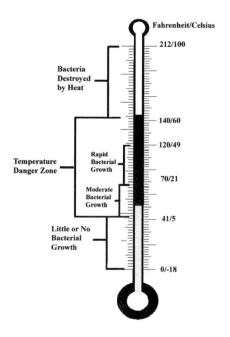

Figure 2.12—Temperature Control Guide

Time

Under ideal conditions, bacterial cells can double in number every 15 to 30 minutes. *Clostridium perfringens* bacteria can double every 10 minutes (Longrèe and Armbruster, 1996). For most bacteria, a single cell can generate over 1 million cells in just 5 hours (Figure 2.13). It is very important not to give bacteria an opportunity to multiply. Proper storage and handling of foods helps to prevent bacteria from multiplying.

Because bacteria have the ability to multiply rapidly, it does not take long before many cells are produced. A rule of thumb in the foodservice industry is that **bacteria need about four hours to grow to high enough numbers to cause illness**. This includes the total time that a food is between 41° and 140°F (5° and 60°C).

Remember, a single bacterial cell can produce over one million cells in just five hours under ideal conditions.

Time	0	15 min.	30 min.	60 min.	3 hrs.	5 hrs.
# of cells	1	2	4	16	>1000	>1 million

Figure 2.13—Bacterial Growth Time

Oxygen

Bacteria also differ in their requirements for oxygen. **Aerobic** bacteria *must* have oxygen in order to grow. **Anaerobic** bacteria, however, cannot survive when oxygen is present because it's toxic to them. Anaerobic bacteria grow well in vacuum packaged foods or canned foods where oxygen is not available. Anaerobic conditions also exist in the middle of cooked food masses such as in large stock pots, baked potatoes, or in the middle of a roast or ham.

Facultative anaerobic forms of bacteria can grow *with* or *without* free oxygen but have a preference. Most foodborne disease-causing microorganisms are facultative anaerobes.

Microaerophilic organisms have a very specific oxygen requirement, usually in the range of three to six percent.

Controlling oxygen conditions may not be an effective way to prevent foodborne illness. Regardless of available oxygen, some disease-causing bacteria will find the conditions suitable for growth.

Moisture

Like most other forms of life, moisture is an important factor in bacterial growth. That is why man has dried foods for thousands of years as a way to preserve them.

Scientists have determined that it is not the percentage of moisture or "water by volume" in a food that most affects bacterial growth.

Rather it is the amount of "available water" or water available for bacterial activity, This is expressed as water activity and is designated with the symbol A_w. **Water activity** is a measure of the amount of water that is not bound to the food and is therefore available for bacterial growth (Figure 2.14).

For example, a fresh chicken has 60% water by volume, and its A_w is approximately .98. The same chicken, when frozen, still has 60% water by volume but its A_w is now "0." Water activity is measured on a scale from 0-1.0. **Disease-causing bacteria can only grow in foods that have a water activity higher than .85.** Many foods are preserved by lowering their water activity to .85 or below. Drying foods or adding salt or sugar reduces the amount of available water.

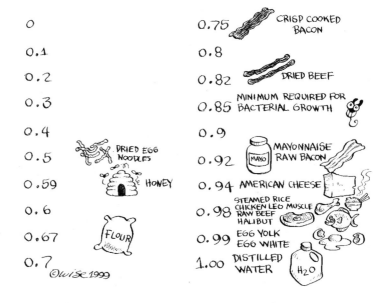

Figure 2.14 Water Activity (A_w) of Some Foods Sold in Food Establishments

Potentially Hazardous Foods (PHF)

Some types of foods have the ability to support the rapid and progressive growth of infectious and toxin-producing microorganisms. These foods are called "potentially hazardous." **Potentially hazardous foods (PHF)** are usually high in protein *and* have a pH above 4.6 *and* a water activity above .85.

Common examples of potentially hazardous foods are red meats, poultry and raw shell eggs, fish and shellfish, and dairy products (Figure 2.15). Other potentially hazardous foods are vegetables such as cooked rice or potatoes, refried beans, and fruits such as cut cantaloupe. The *Food Code* classifies the following as potentially hazardous foods:

▲ Foods of animal-origin

▲ Foods of plant origin that are heat-treated or consist of raw seed sprouts

▲ Cut melons

▲ Garlic and oil mixtures.

Figure 2.15—Examples of Potentially Hazardous Foods

Potentially hazardous foods always require special handling. If these foods are held at temperatures between 41° and 140°F (5° and 60°C) for four hours or more, harmful microorganisms can grow to dangerous levels. Potentially hazardous foods have been associated with most foodborne disease outbreaks. **It is critical that you control the handling and storage of potentially hazardous foods to prevent bacterial growth**.

Bacillus cereus

Campylobacter jejuni

Clostridium perfringens

Clostridium botulinum

Escherichia coli

Listeria monocytogenes

Salmonella spp.

Shigella spp.

Staphylococcus aureus

Vibrio spp.

Hepatitis A

Norwalk virus group

Rotavirus

Anisakis spp.

Cryptosporidium parvum

Cyclospora cayetonensis

Giardia lamblia

Toxoplasma gondii

Trichinella spiralis

Figure 2.16—Common Biological Hazards in Food Establishments

Ready-to-Eat Foods

Ready-to-eat foods can become contaminated if not handled properly. They include raw or processed products that can be eaten immediately. Examples of ready-to-eat foods include:

▲ Delicatessen items such as cheeses and luncheon meats

▲ Vegetables

▲ Salad items

▲ Hot dogs

▲ Hard-boiled eggs.

Foodborne Illness Caused by Bacteria

Bacteria are classified as *sporeforming* and *non-sporeforming* organisms. In the following sections, each type of biological hazard is described, the common foods and route of transmission are identified, and preventive strategies are discussed. In Figure 2.16, common biological hazards are listed.

Foodborne Illness Caused by Sporeforming Bacteria

The following group of bacteria is capable of forming spores. A **spore** structure enables a cell to survive environmental stress such as cooking, freezing, high-salt conditions, drying, and high-acid conditions (Figure 2.17).

Spores are not harmful if ingested, except in a baby's digestive system, where *Clostridium botulinum* spores can cause infant botulism. However, if conditions in the food are suitable for bacterial growth and the spore turns into a vegetative cell, the vegetative cell can grow in the food and cause illness if eaten.

Figure 2.17 Well-Protected Spores

Sporeforming bacteria are generally found in foods that are grown in soil, like vegetables and spices. They may also be found in animal products. They can be particularly troublesome in food establishments when foods are not cooled properly.

For example, a 10-gallon pot of chili was prepared for the next day's lunch. All the ingredients (beans, meat, spices, tomato base) were mixed together and cooked to a rapid boil. Vegetative cells should die, but spores may survive. Refer to Figure 2.6 to see how this happens.

The chili was stored in the 10-gallon pot and allowed to cool overnight in a walk-in refrigerator. It can take the core temperature of the chili 2 to 3 days to cool from 140° to 41°F (60°C to 5°C). If given enough time at the right temperature during the cooling process, sporeforming bacteria that survived the cooking process may change into vegetative cells and begin to grow.

Spores are most likely to turn vegetative when:

▲ Heat-shocked (cooked)

▲ Optimum conditions exist for growth (high protein, high moisture)

▲ Temperatures are in the **food temperature danger zone** or between 41° to 140°F (5° to 60°C)

▲ When the amount of time food is in the danger zone is 4 hours or more.

To prevent sporeforming bacteria from changing to the dangerous vegetative state, it is critical that hot food temperatures be maintained at 140°F (60°C) or above and cold foods should be held at 41°F (5°C) or below. Always cook and cool foods as rapidly as possible to limit bacterial growth.

Bacillus cereus

CAUSATIVE AGENT (*sporeforming bacteria)	TYPE OF ILLNESS	SYMPTOMS ONSET	COMMON FOODS	PREVENTION
*Bacillus cereus	Bacterial intoxication or toxin-mediated infection	1) Diarrhea type: abdominal cramps (8-16 hrs) 2) Vomiting type: vomiting, diarrhea, abdominal cramps (30 min.-6 hrs)	1) Diarrhea type: meats milk vegetables 2) Vomiting type: rice starchy foods grains cereals	Properly heat, cool, and reheat foods

Description—Bacillus cereus is a sporeforming bacterium that can survive with or without oxygen. It has been associated with two very different types of illnesses: one vomiting, the other diarrhea.

Symptoms and Onset Time—Depending on the type of toxin that is produced by the bacteria on the food, the illness will be either a vomiting type or diarrhea type problem. Nausea and abdominal cramps may also be associated with either type. The onset time for the vomiting type disease is 30 minutes to 6 hours, and it usually lasts a day or less. The onset time for the diarrhea type disease is 8 to 16 hours, and it lasts for 12 to 14 hours.

Common Foods—The vomiting type of illness is usually associated with grain products such as rice, potatoes, pasta, corn, cornstarch, soybeans, tofu, and flour. A wide variety of foods, including meats, milk, vegetables, and fish have been associated with the diarrheal type disease.

Transmission in Foods—Illness due to *Bacillus cereus* is most often attributed to foods that are improperly stored (cooled, hot-held), permitting the conversion of spores to vegetative cells. Vegetative cells then produce toxin in the food that leads to illness.

Prevention—Foods must be cooked and, if not consumed immediately, held at 140°F (60°C) or above. Foods must also be cooled rapidly to below 41°F (5°C) prior to storage.

Clostridium perfringens

CAUSATIVE AGENT (*sporeforming bacteria)	TYPE OF ILLNESS	SYMPTOMS ONSET	COMMON FOODS	PREVENTION
Clostridium perfringens	Bacterial toxin-mediated infection	Intense abdominal pains and severe diarrhea, (8-22 hrs)	Spices, gravy, improperly cooled foods (especially meats and gravy dishes)	Properly cook, cool, and reheat foods

Description—*Clostridium perfringens* is a nearly anaerobic (must have very little oxygen), sporeforming bacterium that causes foodborne illness. Perishable foods that have been temperature abused (not kept hot- above 140°F, 60° C; or cold- below 41°F, 5°C) are frequently associated with this problem. *Clostridium perfringens* causes illness due to a toxin-mediated infection where the ingested cells colonize and then produce a toxin in the human intestinal tract.

Symptoms and Onset Time—*Clostridium perfringens* causes abdominal pain and diarrhea. The onset time is 8 to 22 hours and the illness usually lasts one day or less.

Common Foods—The organism is widely distributed and may be found in most foodstuffs. A common source of *Clostridium perfringens* is meat that has been boiled, steamed, braised, stewed, or insufficiently roasted, then allowed to cool slowly and served the next day either cold or improperly reheated.

Transmission in Foods—Illness due to *Clostridium perfringens* is most often attributed to foods that are temperature abused, especially those that have been improperly cooled and reheated.

Prevention—Foods must be cooked to 145°F (63°C) or above. Cooked foods must be cooled from 140° to 70°F (60° to 21°C) within 2 hours and from 70° to 41°F (21° to 5°C) in an additional four hours. Foods must also be reheated to 165°F (74°C) within 2 hours and held at 140°F (60°C) until served. Foods should be reheated only one time. If not consumed after one reheating, the food must be discarded.

Clostridium botulinum

CAUSATIVE AGENT (*sporeforming bacteria)	TYPE OF ILLNESS	SYMPTOMS ONSET	COMMON FOODS	PREVENTION
Clostridium botulinum	Bacterial intoxication	Dizziness, double vision, difficulty in breathing and swallowing, headache (12-36 hrs)	Improperly canned foods, vacuum packed refrigerated foods; cooked foods in anaerobic mass	Properly heat process anaerobically packed foods; DO NOT use home canned foods

Description—*Clostridium botulinum* is an anaerobic (must not have oxygen), sporeforming bacterium that causes foodborne intoxication due to improperly heat-processed foods (especially home-canning). The organism produces a **neurotoxin** which is one of the most deadly biological toxins known to man. This toxin is not heat stable and can be destroyed if the food is boiled for about 20 minutes. However, botulism still occurs because people do not want to boil food that has already been cooked.

Symptoms and Onset Time—Symptoms commonly associated with botulism are fatigue, headache, dizziness, visual disturbances, an inability to swallow, and respiratory paralysis. The onset time is 12 to 36 hours and the duration of the illness is several days to a year.

Common Foods—Low-acid foods (pH of 4.6 or above), that are inadequately heat-processed and then packaged anaerobically (metal can or vacuum pouch), and held in the food temperature danger zone. Examples would include home-canned green beans, meats, fish, and garlic or onions that are stored in oil and butter respectively.

Transmission in Foods—Illness due to *Clostridium botulinum* is almost always attributed to ingestion of foods that were not heat processed correctly and packaged anaerobically.

Prevention—Do not use home-canned foods or buy foods from unapproved sources. Do not can or vacuum package foods in a food establishment. Discard damaged cans (see Chapter 4). The Food Code stipulates that a variance is required for any food establishment that prepares products in reduced-oxygen packaging. *Clostridium botulinum* is not necessarily confined to foods canned or bottled at home. Anaerobic conditions created in hot-food masses which are not cooled properly can become contaminated. For example, a "perpetual" butter stockpot in the corner of the griddle, the advanced cooking of a mass of chopped onions in butter or oil and held on a griddle, or "baked" potatoes left to cool without refrigeration are all capable of producing *Clostridium botulinum* toxin. (Refer to Case Study 2.2 at the end of this chapter.)

Foodborne Illness Caused by Non-Sporeforming Bacteria

The following group of bacteria do not form spores. That is, they stay in the vegetative state all the time. Compared to bacterial spores, vegetative cells are easily destroyed by proper cooking. There are numerous examples of non-sporeforming foodborne bacteria that are important in the food industry.

Campylobacter jejuni

CAUSATIVE AGENT	TYPE OF ILLNESS	SYMPTOMS ONSET	COMMON FOODS	PREVENTION
Campylobacter jejuni	Bacterial infection	Watery, bloody diarrhea (2-5 days)	Raw chicken, raw milk, raw meat	Properly handle and cook foods; avoid cross contamination

Description—*Campylobacter jejuni* is a major cause of foodborne infection. The organism is unique compared to most other foodborne pathogens because it requires a very strict amount of air for growth. As a microaerophile, it can tolerate only 3 to 6 percent oxygen for growth.

Symptoms and Onset Time—*Campylobacter jejuni* commonly causes abdominal pain and slight to severe bloody diarrhea. The onset time is 2 to 5 days. Symptoms of this disease usually last from 2 to 7 days.

Common Foods—This organism is commonly found in raw milk, raw poultry, and raw meats. Some scientists estimate that *Campylobacter jejuni* may be present in nearly 100% of all retail chickens.

Transmission in Foods—*Campylobacter jejuni* is often transferred from raw meats to other foods by cross contamination. This is typically done by transfer from a food contact surface (such as a cutting board or knife) or a food worker's hands.

Prevention—Cook raw meats properly. Thoroughly clean food contact surfaces (cutting boards) and wash hands thoroughly after handling raw foods.

Escherichia coli O157:H7

CAUSATIVE AGENT	TYPE OF ILLNESS	SYMPTOMS ONSET	COMMON FOODS	PREVENTION
Escherichia coli O157:H7	Bacterial infection or toxin-mediated infection	Bloody diarrhea followed by kidney failure and hemolytic uremic syndrome (HUS) in severe cases, (12-72 hrs)	Undercooked hamburger, raw milk, unpasteurized apple cider, lettuce	Practice good food sanitation, handwashing; properly handle and cook foods

Description—The *Escherichia coli* (or *E. coli*) group of bacteria includes four foodborne pathogens; enterotoxigenic *E. coli*, enteropathogenic *E. coli*, enterohemorrhagic *E. coli*, and enteroinvasive *E. coli*. Of particular importance is enterohemorrhagic *E. coli* called *E. coli* O157:H7. This facultative anaerobic bacteria can be found in the intestines of warm-blooded animals, especially cows. The illness caused by *E. coli* can be an infection or a toxin-mediated infection. Only a small amount of bacteria are required to produce an illness.

Symptoms and Onset Time—Illness due to *E. coli* O157:H7 is a threat to children up to 16 years old and the elderly. It can cause severe abdominal pain, nausea, vomiting, bloody diarrhea, kidney failure, and death. The organism burrows into the intestine and produces a toxin which clots the blood (especially in the kidneys) causing Hemolytic Uremic Syndrome (HUS). The infection usually begins with flu-like symptoms and fever followed by bloody diarrhea. The onset time is 12 to 72 hours and the illness usually lasts from 1 to 3 days.

Common Foods—This organism has been isolated in raw milk and in raw and undercooked beef and other red meats, improperly pasteurized milk, raw finfish, and prepared foods. Recent outbreaks were linked to lettuce and unpasteurized apple juice.

Transmission in Foods—*E. coli* O157:H7 is usually transferred to foods like beef through contact with the intestines of slaughter animals. Apples used for juice from orchards where cattle or deer grazed are also suspected. Transmission can occur if employees who are carriers do not wash their soiled hands properly after going to the toilet. Cross contamination by soiled equipment and utensils may also spread *E. coli* O157:H7.

Prevention—Cook ground meats to at least 155°F (68°C). Make sure that employees wash their hands thoroughly before starting to work with food and after going to the toilet. Prevent cross contamination and keep hot foods above 140°F (60°C) and cold foods below 41°F (5°C). Wash lettuce in sinks used only for food preparation. Use only pasteurized apple cider or fruit juice and milk products.

Listeria monocytogenes

CAUSATIVE AGENT	TYPE OF ILLNESS	SYMPTOMS ONSET	COMMON FOODS	PREVENTION
Listeria monocytogenes	Bacterial infection	1) Healthy adult: flu-like symptoms 2) At-risk population: septicemia, meningitis, encephalitis, birth defects 3) Still-birth (1 day- 3 weeks)	Raw milk, dairy items, raw meats, refrigerated ready-to-eat foods, processed ready-to-eat meats such as hot dogs, raw vegetables, and seafood	Properly store and cook foods; avoid cross contamination; rotate processed refrigerated foods using FIFO to ensure timely use

Description—*Listeria monocytogenes* is a facultative anaerobic (can grow with or without oxygen) bacterium that causes foodborne infection. This microbe is important to foodservice operations because it has the ability to survive under many conditions such as in high-salt foods and, unlike most other foodborne pathogens, can grow at refrigerated temperatures (below 41°F; 5°C).

Symptoms and Onset Time—Listeriosis, the illness caused by *Listeria monocytogenes*, usually causes a flu-like disease in healthy adults. Symptoms include nausea, vomiting, headache, fever, chills, and backache. Complications from listeriosis can be life threatening (septicemia, meningitis, encephalitis, birth defects) for pregnant women, or the at-risk population such as the young, elderly, and those with weakened immune systems. The onset time is 1 day to 3 weeks and the duration of the disease is indefinite depending on when treatment is administered to the patient.

Common Foods—This organism is everywhere and has been isolated in many foods. It is most common in raw meats, raw poultry, dairy products (cheeses, ice cream, raw milk), cooked luncheon meats and hot dogs, raw vegetables, and seafood.

Transmission in Foods—Transmission to foods can occur by cross contamination or if foods are not cooked properly.

Prevention—Cook foods thoroughly and practice good food handling techniques. Specific practices for control are timely use and rotation of refrigerated products, such as cooked turkey breasts, hot dogs, hams, and luncheon meats.

Salmonella spp. (*spp.* means species of)

CAUSATIVE AGENT	TYPE OF ILLNESS	SYMPTOMS ONSET	COMMON FOODS	PREVENTION
Salmonella spp.	Bacterial infection	Nausea, fever, vomiting, abdominal cramps, diarrhea (6-48 hrs)	Raw meats, raw poultry, eggs, milk, dairy products	Properly cook foods; avoid cross contamination

Description—*Salmonella* are facultative anaerobic (grow with or without oxygen) bacteria frequently implicated as a foodborne infection. *Salmonella* is found in the intestinal tract of humans and warm-blooded animals. It frequently gets into foods as a result of fecal contamination.

Symptoms and Onset Time—*Salmonella* infection produces abdominal pain, headache, nausea, vomiting, fever, and diarrhea. The onset time is 6 to 48 hours and the disease generally lasts 2-3 days.

Common Foods—This organism is inherent to many foods, especially raw meat and poultry products, eggs, milk, dairy products, pork, chocolate, and cream-filled desserts.

Transmission in Foods—Transmission to foods is commonly through cross contamination where fecal material is transferred to food through contact with raw foods (especially poultry), contaminated food contact surfaces (i.e., cutting boards), or infected food workers. The incidence of salmonellosis has risen more than twenty-fold since 1946. One reason may be the increased number of centrally processed foods along with bulk distribution (Longrèe and Armbruster, 1996).

Prevention—Cook foods thoroughly. Clean and sanitize raw food contact surfaces after use; and make sure food workers wash their hands adequately before working with food, especially after going to the toilet.

Shigella spp.

CAUSATIVE AGENT	TYPE OF ILLNESS	SYMPTOMS ONSET	COMMON FOODS	PREVENTION
Shigella spp.	Bacterial infection	Bacillary dysentery, diarrhea, fever, abdominal cramps, dehydration, (1-7 days)	Foods that are prepared with human contact: salads, raw vegetables, milk, dairy products, raw poultry, non-potable water, ready to eat meat	Wash hands and practice good personal hygiene; properly cook foods

Description—*Shigella* are facultative anaerobic bacteria that account for about 10 percent of foodborne illnesses in the United States. These organisms are commonly found in the intestines and feces of humans and warm-blooded animals. They cause shigellosis, a foodborne infection. The bacterium produces a toxin that reverses the absorption of water back into the body. Water flows into the gut and feces to cause watery diarrhea.

Symptoms and Onset Time—Shigellosis causes diarrhea, fever, abdominal cramps, chills, fatigue, and dehydration. The onset time is 1 to 7 days and the duration of the disease frequently depends on when treatment is administered to the patient.

Common Foods—This organism is common in ready-to-eat salads (i.e., potato, chicken), milk and dairy products, poultry, raw vegetables, and any food contaminated by feces that contain the microbe.

Transmission in Foods—Water that is contaminated by fecal material and food and utensils handled by workers who are carriers of the bacteria can cause this problem.

Prevention—Do not allow individuals who have been diagnosed with shigellosis to handle food. Make certain that food workers wash their hands thoroughly after going to the toilet. Cook foods to proper temperature and do not allow cross contamination. Wash produce and other foods with potable water (water that is safe to drink).

Staphylococcus aureus

CAUSATIVE AGENT	TYPE OF ILLNESS	SYMPTOMS ONSET	COMMON FOODS	PREVENTION
Staphylococcus aureus	Bacterial intoxication	Nausea, vomiting, abdominal cramps, headaches (2-6 hrs)	Foods that are prepared with human contact, cooked or processed foods	Wash hands and practice good personal hygiene. Cooking WILL NOT inactivate the toxin

Description—*Staphylococcus aureus* is a facultative anaerobic bacterium that produces a heat stable toxin as it grows on foods. This bacterium can also grow on cooked, and otherwise safe, foods that are re-contaminated by food workers who mishandle the food. *Staphylococcus aureus* bacteria do not compete well when other types of microorganisms are present. However, they grow well when alone and without competition from other microbes. *Staphylococcus aureus* bacteria produce toxins that cause an intoxication. These bacteria are commonly found on human skin, hands, hair, and in the nose and throat. They may also be found in burns, infected cuts and wounds, pimples, and boils. At any given time, *Staphylococcus aureus* bacteria can be found living in and on 30 to 50% of people, including those who are otherwise healthy. These organisms can be transferred to foods easily, and they can grow in foods that contain high salt or high sugar, and a lower water activity.

Symptoms and Onset Time—*Staphylococcus aureus* intoxication produces severe nausea, acute abdominal pain, vomiting and diarrhea. The onset time is 1 to 6 hours, usually 2 to 4 hours. The illness lasts 1-2 days.

Common Foods—This organism has been found in cooked ready-to-eat foods such as luncheon meats, ready-to-eat meat, vegetable and egg salads (like taco, potato, and tuna salad), meat, poultry, custards, high-salt foods like ham, and milk and dairy products.

Transmission in Foods—Since humans are the primary reservoir, contamination from the worker's hands is the most common way the organism is introduced into foods. Foods requiring considerable food preparation and handling are especially susceptible. The bacteria are also spread by droplets of saliva from talking, coughing, and sneezing near food. Food workers who improperly use tasting spoons and ladles can transfer bacteria from their mouth to food.

Prevention—Avoid contamination of food from bare hands. Do not allow individuals who have infected cuts, burns, or wounds to handle food unless the wound has been properly bandaged and a plastic glove or similar barrier is used to cover the bandaged area. Make certain that food workers wash their hands thoroughly before starting to work and whenever their hands become contaminated. Do not allow food workers to re-use tasting spoons and ladles. Heat and cool foods properly, and keep ready-to-eat foods out of the temperature danger zone.

Vibrio spp.

CAUSATIVE AGENT	TYPE OF ILLNESS	SYMPTOMS ONSET	COMMON FOODS	PREVENTION
Vibrio spp.	Bacterial infection	Headache, fever, chills, diarrhea, vomiting, severe electrolyte loss, gastroenteritis, (2-48 hrs)	Raw or improperly cooked fish and shellfish	Practice good sanitation; properly cook foods; avoid serving raw seafood

Description—There are three organisms within the *Vibrio* group of bacteria that have been connected with foodborne infections. They include *Vibrio cholera*, *Vibrio parahaemolyticus*, and *Vibrio vulnificus*. All are important since they are very resistant to salt and are common in seafood.

Symptoms and Onset Time—Foodborne illness from *Vibrio spp.* is characterized by diarrhea, abdominal cramps, nausea, vomiting, headache, fever, and chills. The onset time is 2 to 48 hours, and the disease typically lasts 2 to 3 days, but can be longer.

Common Foods—*Vibrio spp.* are commonly found in raw, under-processed, improperly handled, and contaminated fish and shellfish. Common vehicles of the bacteria are clams, oysters, crabs, shrimp, and lobster. This bacteria is more common in summer months and in seafood harvested from warmer waters. *Vibrio cholera* is spread by fecal contamination, especially in seafood that is harvested from polluted waters; *Vibrio vulnificus* bacteria are natural to certain saltwater species and can kill individuals in the at-risk population who have damaged livers. *Vibrio parahoemolyticus* is natural in certain species of saltwater fish and seafood.

Transmission in Foods—Since the organism is inherent in many raw seafoods, transmission to other foods by cross contamination is a concern. Most illnesses are caused due to the consumption of raw or undercooked seafood.

Prevention—Buy seafood from approved sources only. Cook seafood to the proper temperature, and avoid eating raw or lightly-cooked seafood, especially if you are in the at-risk segment of the population. Food workers must practice good personal hygiene and use correct handwashing procedures.

Foodborne Illness Caused by Viruses

The viruses that cause foodborne disease differ from foodborne bacteria in several ways. **Viruses** are much smaller than bacteria, and they require a living host (human, animal) to grow and reproduce. Viruses do not multiply in foods. However, a susceptible person needs to consume only a few viral particles in order to experience an infection.

Viruses are usually transferred from one food to another, from a food worker to a food, or from a contaminated water supply to a food. **A potentially hazardous food is not needed to support survival of viruses.** There are three viruses that are of primary importance to food establishments: Hepatitis A, Norwalk, and Rotavirus. **Proper handwashing, especially after using the toilet, is the key to controlling the spread of foodborne viruses**.

Hepatitis A

CAUSATIVE AGENT	TYPE OF ILLNESS	SYMPTOMS ONSET	COMMON FOODS	PREVENTION
Hepatitis A	Viral infection	Fever, nausea, vomiting, abdominal pain, fatigue, swelling of the liver, jaundice (15-50 days)	Foods that are prepared with human contact; contaminated water	Wash hands and practice good personal hygiene; avoid raw seafood

Description—Hepatitis A is a foodborne virus that has been associated with many foodborne infections. Hepatitis A causes a liver disease called infectious hepatitis. The hepatitis virus is a particularly important hazard to food establishments because food workers can harbor it for up to 6 weeks and not show symptoms of illness. Food workers are contagious for one week before onset of symptoms and two weeks after the symptoms of the disease appear. During that time, infected workers can contaminate foods and other workers by spreading fecal material from unwashed hand and nails. Hepatitis A virus is very hardy and can live for several hours in a suitable environment.

Symptoms and Onset Time—The symptoms of infectious hepatitis are fever, nausea, vomiting, abdominal pain, and fatigue. Advanced stages of the disease cause swelling of the liver and possibly jaundice—a yellowing of the skin. The onset time for infectious hepatitis is 15 to 50 days after eating the contaminated food. A mild case of infectious hepatitis typically lasts several weeks. More severe cases can last several months.

Common Foods—Hepatitis A virus can be found in raw and lightly-cooked oysters and clams that are harvested from polluted waters. Also, raw vegetables that have been irrigated or washed with polluted water can be contaminated. Most potentially hazardous foods, if they are mishandled by a food worker who is infected with the Hepatitis A virus, can carry the microbe. Examples of these foods include prepared salads, sliced luncheon meats, salad-bar items, sandwiches, and bakery products. Due to the onset time, it is very difficult to identify the food source of Hepatitis A infection.

Transmission in Foods—The virus is transmitted by ingestion of food and water that contain the Hepatitis A virus.

Prevention—Handle foods properly and cook them to recommended temperatures. Avoid eating raw seafood. Food workers must practice good personal hygiene and wash hands and fingernails thoroughly before working with food and after going to the toilet.

Norwalk virus

CAUSATIVE AGENT	TYPE OF ILLNESS	SYMPTOMS ONSET	COMMON FOODS	PREVENTION
Norwalk virus	Viral infection	Vomiting, diarrhea, abdominal pain, headache, low grade fever; onset 24-48 hrs	Sewage, contaminated water; contaminated salad ingredient, raw clams, oysters, and infected food workers	Use potable water; cook all shellfish; handle food properly, meet time, temperature guidelines for PHF

Description—The Norwalk virus is another common foodborne virus that has been associated with many foodborne infections.

Symptoms and Onset Time—Common symptoms caused by Norwalk and Norwalk-like viruses are nausea, vomiting, diarrhea, and abdominal pain. Headache and low-grade fever may occur. A mild, brief illness usually develops 24 to 48 hours after contaminated food or water is consumed and lasts about one to three days. Severe illness is very rare.

Common Foods—Sewage-contaminated water is the most common source of Norwalk outbreaks. Shellfish and salad ingredients are the foods most often associated with this illness. Ingestion of raw or insufficiently-steamed clams and oysters poses a high risk for infection with Norwalk virus. Foods other than shellfish are contaminated by ill food workers.

Transmission in Foods—The virus is primarily transmitted by ingestion of food and water that has been contaminated with feces that contain the Norwalk virus.

Prevention—Use potable water (safe to drink) when preparing shellfish. Handle foods properly and cook foods to the correct temperature. Avoid consumption of raw seafood. Food workers must practice good personal hygiene and wash hands and fingernails thoroughly before working with food and after going to the toilet.

Rotavirus

CAUSATIVE AGENT	TYPE OF ILLNESS	SYMPTOMS ONSET	COMMON FOODS	PREVENTION
Rotavirus	Viral infection	Diarrhea (especially in infants and children), vomiting, low grade fever; 1-3 days onset; lasts 4-8 days	Sewage, contaminated water, contaminated salad ingredients, raw seafood	Good personal hygiene and handwashing; proper food handling practices

Description—Group A rotaviruses cause several diseases known as rotavirus gastroenteritis. Group A rotavirus is the leading cause of severe diarrhea among infants and children. Over 3 million cases of rotavirus gastroenteritis occur annually in the United States.

Symptoms and Onset Time—Individuals infected with rotavirus may experience mild to severe symptoms. Common symptoms of rotavirus gastroenteritis are vomiting, watery diarrhea, and low-grade fever. The onset time ranges from 1-3 days. Symptoms often start with vomiting followed by 4-8 days of diarrhea.

Common Foods—Infected food workers may contaminate foods that require handling and no further cooking, such as salads and fruits.

Transmission in Foods—Person-to-person spread through contaminated hands is probably the most common means by which rotaviruses are transmitted in day care centers and family homes. Contaminated hands may spread the virus to foods that will not be cooked before eating.

Prevention—Cook foods to the proper temperature and handle properly. Food workers must practice good personal hygiene and wash hands and fingernails thoroughly before working with food and after going to the toilet.

Foodborne Illness Caused by Parasites

Foodborne parasites are another important foodborne biological hazard. **Parasites** are small or microscopic creatures that need to live on or inside a living host to survive. There are many parasites that can enter the food system and cause foodborne illness. In this chapter, we list a few of the most troublesome ones that may appear in food establishments. Parasitic infection is far less common than bacterial or viral foodborne illnesses.

Anisakis spp.

CAUSATIVE AGENT	TYPE OF ILLNESS	SYMPTOMS ONSET	COMMON FOODS	PREVENTION
Anisakis spp.	Parasitic infection	Coughing, vomiting onset 1 hour to 2 weeks	Raw or undercooked seafood-especially bottom feeding fish	Cook fish to the proper temperature throughout; freeze to meet Food Code specifications

Description—*Anisakis spp.* are nematodes (roundworms) associated with foodborne infection from fish. The worms are about 1–1-1/2 inches long and the diameter of a human hair. They are beige, ivory, white, gray, brown, or pink. Other names for this parasite are "cod worm" (not to be confused with common roundworms found in cod) and "herring worm."

Symptoms and Onset Time—If the worms attach themselves to the victim's stomach, the typical symptoms are vomiting and abdominal pain. Coughing is the most common symptom if the worms attach themselves in the throat. When the worms attach themselves in the large intestine they produce sharp pain and fever, symptoms similar to those produced by appendicitis.

Common Foods—*Anisakis spp.* have been implicated in foodborne disease after consumption of raw or undercooked seafood. This organism has been found in bottom-feeding fish such as cod, haddock, fluke, salmon, herring, flounder, monkfish, plus sea urchins, crab, shrimp, starfish, and tuna. Cerviche or fresh seafood salad are known to carry this parasite.

Transmission in Foods—The natural hosts of the parasite are walruses, and perhaps sea lions and otters. The worms are transferred to fish, their intermediate host, in the water in which the walruses live. Humans become the accidental host upon eating fish that are infested with the parasites. Humans do not make good hosts for the parasites. The worms will not complete their life cycles in humans, and eventually die.

Prevention—Inspect seafood and handle it carefully to expose worms. Cook seafood to proper temperature.

Cyclospora cayetanensis

CAUSATIVE AGENT	TYPE OF ILLNESS	SYMPTOMS ONSET	COMMON FOODS	PREVENTION
Cyclospora cayentanensis	Parasitic infection	Watery and explosive diarrhea, loss of appetite, bloating (1 week)	Water, strawberries, raspberries, and raw vegetables	Good sanitation, reputable supplier

Description—*Cyclospora cayentanensis* is a parasite that has been reported much more frequently beginning in the 1990s. *Cyclospora* frequently finds its way into water and then can be transferred to foods. It can also be transferred to foods during handling. The most recent outbreaks of cyclosporiasis have been associated with fresh fruits and vegetables that were contaminated at the farm.

Symptoms and Onset Time—Cyclosporiasis is an infection that acts upon the small intestine usually resulting in watery (and sometimes explosive) diarrhea. Other symptoms include loss of appetite, weight loss, bloating, stomach cramps, nausea, vomiting, muscle aches, low-grade fever, and fatigue. Symptoms usually start within one week after consuming the food. If the illness goes untreated, symptoms can persist for many weeks or months.

Common Foods—Recent cases of cyclosporiasis have been attributed to contaminated water, raspberries, strawberries, and other kinds of fresh produce.

Transmission in Foods—*Cyclospora* is passed from person to person by fecal-oral transmission. Foods usually become contaminated after coming in contact with fecal material from polluted water or a contaminated food worker. The *Cyclospora* parasite may take days or weeks after a person eats a contaminated food to become infectious.

Prevention—Prevention against *Cyclospora* is similar to other parasites and viruses. Avoid contact with contaminated foods, water, or food workers. Good sanitation and purchasing ready-to-eat foods from an inspected and approved supplier are also very important.

Cryptosporidium parvum, Giardia lamblia

CAUSATIVE AGENT	TYPE OF ILLNESS	SYMPTOMS ONSET	COMMON FOODS	PREVENTION
Cryptosporidium parvum	Parasitic infection	Severe watery diarrhea within 1 week of ingestion	Contaminated water; food contaminated by infected food handlers	Use potable water supply; practice good personal hygiene and handwashing
Giardia lamblia	Parasitic infection	Diarrhea within 1 week of contact	Contaminated water	Potable water supply; good personal hygiene and handwashing

Description—*Crytosporidium parvum* and *Giardia lamblia* are single-cell microorganisms called protozoa. *Cryptosporidium* is found in water that has been contaminated with cow feces. *Giardia* is found in the feces of wild animals, domestic pets, and infected persons. Both types of organisms can cause foodborne infection. These organisms are important sources of non-bacterial diarrhea in the United States.

Symptoms and Onset Time—The most common symptom of intestinal cryptosporidiosis is severe watery diarrhea, which usually lasts 2 to 4 days. Human giardiasis may cause diarrhea within one week of ingestion of the cyst. The illness may last as much as a month. The disease is more severe among the at-risk population.

Common Foods—These parasites are most commonly found in water and raw foods that have been polluted with sewage containing the cysts of *Cryptosporidium* or *Giardia. Cryptosporidium* could occur on any food touched by a contaminated food worker. Incidence is higher in child care centers that serve food. Fertilizing salad vegetables with manure is another possible source of human infection. Five outbreaks of giardiasis have been traced to food contamination by infected food workers. Infections from contaminated vegetables that are eaten raw cannot be ruled out. Cool moist conditions favor the survival of the organism.

Transmission in Foods—These parasites could occur, theoretically, on any food touched by a contaminated food worker. They are primarily transmitted by a water supply that is contaminated with feces and by fecal contamination of food and food contact surfaces.

Prevention—Provide a potable water supply in the food establishment and handle foods carefully to prevent contamination and cross contamination. Food workers must practice good personal hygiene and wash hands thoroughly before working with food and after going to the toilet.

Taxoplasma gondii

CAUSATIVE AGENT	TYPE OF ILLNESS	SYMPTOMS ONSET	COMMON FOODS	PREVENTION
Taxoplasma gondii	Parasitic infection	Mild cases of the disease involve swollen lymph glands, fever, headache, and muscle aches. Severe cases may result in damage to the eye or the brain (10-13 days)	Raw meats, raw vegetables, and fruit	Good sanitation, reputable supplier, proper cooking

Description—Toxoplasmosis is a parasitic infection caused by *Toxoplasma gondii*. The parasite is common in warm-blooded animals including cats, rats, mice, pigs, cows, sheep, chickens, and birds.

Symptoms and Onset Time—Healthy children and adults usually do not experience any symptoms when infected with Toxoplasmosis. Symptoms of mild cases of the illness include swollen lymph glands, fever, headache, and muscle aches. People with immune system problems and persons who have recently received an organ transplant may develop severe toxoplasmosis that results in damage to the eye or the brain. Babies who become infected before birth can be born with mental retardation, blindness, or other serious mental or physical problems. The onset time is 10-13 days after exposure.

Common Foods—The parasites are found in red meat, especially pork, lamb, venison, and beef. Fruits and vegetables can be contaminated with feces.

Transmission in Foods—The domestic cat appears to be a major culprit in transmitting the parasite to humans and other animals. Food can be contaminated by contact with fecal material. Humans most often acquire infections by ingesting cysts in undercooked red meat. Unborn babies can catch this parasite from their mother if the mother is infected during pregnancy.

Prevention—Avoid eating raw and undercooked meat. Wash fruits and vegetables before eating, and wash hands after handling raw meat and vegetables. Prevent cross-contamination from raw foods to cooked or ready-to-eat foods by washing and sanitizing equipment and utensils.

Trichinella spiralis

CAUSATIVE AGENT	TYPE OF ILLNESS	SYMPTOMS ONSET	COMMON FOODS	PREVENTION
Trichinella spiralis	Parasitic infection from a nematode worm	Nausea, vomiting, diarrhea, sweating, muscle soreness (2-28 days)	Primarily undercooked pork products and wild game meats (bear, walrus)	Cook foods to the proper temperature throughout

Description—*Trichinella spiralis* is a foodborne roundworm that causes a parasitic infection. It must be eaten with the infected fleshy muscle of certain meat-eating animals to be transmitted to a new host.

Symptoms and Onset Time—The first symptoms of trichinosis are nausea, vomiting, diarrhea, and abdominal pain. Later stages of the disease are characterized by fever, swelling of tissues around the eyes, and muscle stiffness. The onset time is 2 to 28 days after eating the contaminated meat. Death may occur in severe cases.

Common Foods—Pork is by far the most common vehicle for *Trichinella spiralis*. However, it can also be found in wild-game animals, such as bear, walrus, and wild boar.

Transmission in Foods—This parasite is frequently carried by meat-eating, scavenger animals. These animals are exposed to the parasite when they eat infected tissues from other animals, and garbage that contains contaminated raw-meat scraps.

Prevention—Cook pork and wild-game animals to the proper temperature. Make certain there are no signs of pink color in cooked pork.

Foodborne Illness Caused by Chemicals

Chemical hazards are usually classified as either naturally occurring or man-made chemicals. Naturally occurring chemicals include toxins that are produced by a biological organism. Man-made chemicals include substances that are added, intentionally or accidentally, to a food during processing. Figure 2.18 provides a summary of some of the more common naturally occurring and man-made chemicals.

Naturally Occurring:	Ciguatoxin
	Mycotoxin
	Scombrotoxin
	Shellfish toxins
Man-Made Chemicals:	Cleaning solutions
	Food additives
	Pesticides
	Heavy metals

Figure 2.18—Types of Chemical Hazards in a Food Establishment
(Source: *Food Code*)

Naturally Occurring Chemicals

Ciguatoxin

CAUSATIVE AGENT	TYPE OF ILLNESS	SYMPTOMS ONSET	COMMON FOODS	PREVENTION
Ciguatoxin	Fish toxin originating from toxic algae of tropical waters	Vertigo, hot/cold flashes, diarrhea, vomiting (15 min -24 hours)	Marine finfish including grouper, barracuda, snapper, jack, mackerel, triggerfish, reef fish	Purchase fish from a reputable supplier; cooking WILL NOT inactivate the toxin

Description—Ciguatoxin poisoning is an example of an intoxication caused by eating contaminated tropical reef fish. The toxin is found in tiny, free-swimming sea creatures called algae which live among certain coral reefs. When the toxic algae are eaten by small reef fish, it is stored in the flesh, skin, and organs. When the small reef fish are eaten by bigger fish such as barracuda, mackerel, mahi, bonito, jackfish, and snapper, the toxin accumulates in the flesh and skin of the consuming fish. The contaminated fish are not affected by the toxin. The toxin is heat stable and not destroyed by cooking. At the present time, there is no commercially known method to determine if ciguatoxin is present in a particular fish.

Symptoms and Onset Time—The symptoms of the disease ciguatera vary a lot but include nausea, vomiting, diarrhea, dizziness, and shortness of breath. The classic symptom of ciguatera is a reversal of hot and cold sensations. The first symptoms usually appear within 30 minutes to 6 hours after the contaminated fish is eaten. Most symptoms stop within a few days, but complete recovery may take several weeks or months. Death can occur from a concentrated dose of the toxin.

Common Foods—Marine finfish are most commonly implicated with ciguatoxin poisoning. Common marine species include barracuda, grouper, jack, mackerel, snapper, and triggerfish.

Transmission in Foods—The toxin is transferred to finfish when they eat toxin-containing algae or other fish that contain the toxin.

Prevention—The toxin is not destroyed by cooking; therefore, prevention can be very difficult. Purchasing seafood from a reputable supplier is the best preventive measure.

Scombrotoxin

CAUSATIVE AGENT	TYPE OF ILLNESS	SYMPTOMS ONSET	COMMON FOODS	PREVENTION
Scombrotoxin	Seafood toxin originating from histamine producing bacteria	Dizziness, burning feeling in the mouth, facial rash or hives, peppery taste in mouth, headache, itching, teary eyes, runny nose (1 - 30 min)	Tuna, mahi-mahi, bluefish, sardines, mackerel, anchovies, amberjack, abalone	Purchase fish from a reputable supplier; store fish at low temperatures to prevent growth of histamine-producing bacteria; toxin IS NOT inactivated by cooking

Description—Scombrotoxin, also called histamine poisoning, is caused by eating foods high in a chemical compound called histamine. Histamine is usually produced by certain bacteria when they decompose foods containing the protein histidine. Dark meat of fish has more histidine than other fish meat. Histamine is not inactivated by cooking.

Symptoms and Onset Time—Symptoms of the illness include dizziness, a burning sensation, facial rash or hives, shortness of breath, and a peppery taste in the mouth when contaminated fish is eaten. The symptoms of scombroid poisoning begin within a few minutes to one-half hour after the toxic fish is eaten. The illness lasts only a short period of time, and recovery usually occurs within 8 to 12 hours. In severe cases, loss of muscle control, inability to speak, swallow, or breathe can lead to death (Benenson et al., 1995).

Common Foods—The most common foods implicated in scombrotoxin include tuna, anchovies, blue fish, mackeral, amberjack, abalone, and mahi-mahi, the dark meat fishes. Swiss cheese has also been implicated.

Transmission in Foods—Over time, bacteria that is inherent to a particular food can break down histidine and cause the production of histamine. Leaving fish out at room temperature usually results in histamine production.

Prevention—Purchase seafood from a reputable supplier. Store "fresh" seafood at temperatures between 32° and 39°F (0° and 4°C), and do not accept seafood that is suspected of being thawed and refrozen or temperature abused in some way.

Shellfish Toxins—PSP, DSP, DAP, NSP

CAUSATIVE AGENT	TYPE OF ILLNESS	SYMPTOMS ONSET	COMMON FOODS	PREVENTION
Shellfish toxins: PSP, DSP, DAP, NSP	Intoxication	Numbness of lips, tongue, arms, legs, neck; lack of muscle coordination (10 to 60 min.)	Contaminated mussels, clams, oysters, scallops	Purchase from a reputable supplier

Description—There are several shellfish toxins that have been associated with foodborne illness. The most common illnesses are

Paralytic Shellfish Poisoning (PSP), Diarrhetic Shellfish Poisoning (DSP), Domoic Acid Poisoning (DAP), and Neurotoxic Shellfish Poisoning (NSP). The toxins are produced by certain algae called dinoflagellates. When filter-feeding shellfish, such as mussels, clams, oysters, and scallops, feed on the toxic algae, they accumulate the toxins in their internal organs and become toxic to humans. The amount of toxin in the shellfish depends on the amount of toxic algae (dinoflagellates) in the water and the amount of water filtered by the shellfish.

Symptoms and Onset Time—The first symptoms of the disease are numbness in the lips, tongue, and tips of the fingers. As the disease progresses, symptoms include numbness in the arms, legs, and neck and a lack of muscle coordination. Severe respiratory distress and muscular paralysis accompanies advanced stages of the disease. The symptoms of shellfish poisoning usually occur within 30 to 60 minutes after eating the toxic shellfish. In the case of Domoic Acid poisoning, symptoms can begin as early as 10 minutes after ingestion.

Common Foods—PSP is more common with mussels, clams, oysters, and scallops. DSP is more common with mussels, oysters, and scallops; DAP is more common with mussels in cold waters; NSP is common for Gulf Coast marine (saltwater) animals.

Transmission in Foods—Most cases of seafood toxins are caused by contaminated shellfish that have been harvested by sport fishermen or poached from polluted waters. Commercially harvested shellfish are rarely involved in foodborne disease outbreaks because health agencies monitor the level of toxin in shellfish during high risk periods (May to October). Toxin levels in the shellfish are determined in public health laboratories. When the amount of toxin in the shellfish exceeds safe limits, sale of the shellfish from that harvesting area is prohibited.

Prevention—Purchase shellfish only from a reputable supplier. Avoid buying shellfish that have been harvested by sport fishermen or poached from polluted waters.

Mycotoxins

CAUSATIVE AGENT	TYPE OF ILLNESS	SYMPTOMS ONSET	COMMON FOODS	PREVENTION
Mycotoxins	Intoxication	1)Acute onset-hemorrhage, fluid buildup 2) Chronic onset-cancer from small doses over time	Moldy grains, corn, corn products, peanuts, pecans, walnuts, and milk	Purchase food from a reputable supplier; keep grains and nuts dry; and protect products from humidity

Description—Mycotic organisms or fungi are molds, yeasts, and mushrooms, some of which are capable of causing foodborne illness. Fungi are larger than bacteria, and they usually prefer foods that are high in sugar or starches. Molds and yeasts withstand more extreme conditions (more acidic foods, lower A_w foods) than bacteria can. Most molds and yeasts are spoilage organisms which cause foods to deteriorate. However, some types of fungi produce toxic chemicals called mycotoxins. Many mycotoxins have been shown to cause cancer. An important foodborne mycotoxin, called aflatoxin, is produced by *Aspergillus spp.* molds. Many mycotoxins are not destroyed by cooking.

Symptoms and Onset Time—Acute aflatoxicosis is produced when moderate to high levels of aflatoxins are consumed. Acute episodes of the disease may include hemorrhage, acute liver damage, fluid buildup in the body, altered digestion and metabolism of nutrients, and possibly death. The main concern with mycotoxins is that they can cause cancer if consumed over a long period of time.

Common Foods—Mycotoxins are commonly found in drier or more acidic foods that do not support the growth of bacteria. Common foods containing mycotoxins include corn and corn products, peanuts, pecans, walnuts, and milk.

Transmission in Foods—Foodborne fungi are important because they can produce chemical compounds called mycotoxins which have been linked to cancer.

Prevention—Purchase food from a reputable supplier. Keep grains and nuts dry and protected from humidity.

Added Man-Made Chemicals

There is an extensive list of chemicals added to foods that may pose a potential health risk. Intentionally added chemicals may include food additives, food preservatives, and pesticides. Pesticides leave residues on fruits and vegetables, and can usually be removed by a vigorous washing procedure. Non-intentionally added chemicals may include contamination by chemicals such as cleaning and sanitary supplies. Also, chemicals from containers or food contact surfaces of inferior metal that are misused may lead to heavy-metal or inferior-metal poisoning (cadmium, copper, lead, galvanized metals, etc.).

Food Allergens

Between 5 and 8 percent of children and 1 to 2 percent of adults are allergic to certain chemicals in foods and food ingredients. These chemicals are commonly referred to as food allergens. A **food allergen** causes a person's immune system to "overreact." Some common symptoms of food allergies are hives; swelling of the lips, tongue, and mouth; difficulty breathing or wheezing; and vomiting, diarrhea, cramps. These symptoms can occur in as little as 5 minutes. In severe situations, a life threatening allergic reaction called **anaphylaxis** can occur. Anaphylaxis is a condition that occurs when many parts of the body become involved in the allergic reaction. Symptoms of anaphylaxis include itching and hives; swelling of the throat and difficulty breathing; lowered blood pressure and unconsciousness.

About 90% of all allergies are caused by eight foods (Figure 2.19). **The only way for a person who is allergic to one of these foods to avoid an allergic reaction is to avoid the food. In many cases, it doesn't take much of the food to produce a severe reaction.** As little as half a peanut can cause a severe reaction in highly sensitive people.

Allergies can be very serious. **You need to know which foods in your establishment contain these ingredients.** The FDA requires ingredients to be listed on the label of packaged foods. Always read label information to determine if food allergens may be present.

▲ Milk	▲ Soy
▲ Egg	▲ Tree nuts
▲ Wheat Proteins	▲ Fish
▲ Peanuts	▲ Shellfish

Figure 2-19—Common Food Allergens

Foodborne Illness Caused by Physical Hazards

Physical hazards are foreign objects in food that can cause illness and injury (Figure 2.20). They include items such as fragments of glass from broken glasses, metal shavings from dull can openers, unfrilled toothpicks that may contaminate sandwiches, human hair and jewelry, or bandages that may accidentally be lost by a food handler and enter food. Stones, rocks, or wood particles may contaminate raw fruits and vegetables, rice, beans, and other grain products.

Physical hazards commonly result from accidental contamination and poor food handling practices that can occur at various points in the food chain from harvest to consumer. To prevent physical hazards, wash raw fruits and vegetables thoroughly and visually inspect foods that cannot be washed (such as ground beef). Food workers must be taught to handle food safely to prevent contamination by unwanted foreign objects such as glass fragments and metal shavings. Finally, food workers should not wear jewelry when involved in the production of food, except for a plain wedding band.

Figure 2.20—Common Physical Hazards in a Food Establishment
(Source: *Food Code*)

In this chapter you learned about the many agents that can cause foodborne illness. The summary of disease causing agents presented in Appendix B provides a quick reference to foodborne illnesses. In the busy work world of a food manager, this reference can save you time when searching for such information.

Summary

The case study at the beginning of this chapter illustrates a number of very important points for retail food safety. First, never use homemade food items unless the home's kitchen has been inspected and approved by the proper regulatory authority in the jurisdiction. Second, keep toxic chemicals in their original container or a container that is clearly and correctly labeled. Third, keep toxic substances apart from food during storage. Nonprofessional pesticide users and certified operators should be alert to the acute toxicity these types of pesticides can have when they are accidentally or intentionally introduced into food.

There are many foodborne hazards that a food establishment may encounter. These hazards differ depending on the type of food and method of preparation involved. Food establishments are typically toward the end of the food production chain because foods are prepared, served, and eaten or transported from there. Therefore, it is very important to control and prevent foodborne hazards that could lead to foodborne illness. Control and prevention of foodborne hazards in a food establishment start with understanding the different types of foodborne hazards. The next step is to understand how to control foodborne hazards with good personal hygiene, prevention of cross contamination, as well as proper storage, cooking, cooling, and reheating of foods. These items will be covered in the next chapter.

Case Study 2.1

Bruce, the morning prep cook at a local restaurant, was preparing shredded cheese to be used on pizza. His procedure included cutting the cheese into small blocks and then shredding the cheese by hand. He prepared several 4-gallon containers and left the containers out at room temperature (70°F, 21°C) until use. Three of the four containers were used on pizzas later that day. The next day, the fourth container of cheese was used. On both days, the pizzas were cooked in an oven set to 500°F (260°C).

Four days later, several people came back to the restaurant and said that they had become ill a few hours after eating pizza there. Only people who ate pizza on the second day appeared to become ill.

1. What foodborne hazard may have been associated with this foodborne illness?

2. How could this have been prevented?

Case Study 2.2

In Albuquerque, New Mexico, a well-known food establishment, identified as Restaurant A, had been packing customers in for years. The owners took pride in keeping the establishment spotlessly clean and serving daily thousands of thick juicy steaks and gigantic Idaho potatoes.

When some baked potatoes were left at closing time, they were stored on a kitchen counter or shelf overnight. The next morning, the salad chefs would arrive to peel, dice, and mix the leftover potatoes with other ingredients to make the side order special of the day: Potato Salad. This had been a standard practice for years. The finished product was containerized and chilled before serving at noon.

One day, the potato salad was made and served as usual. Customers came in for lunch and some ate potato salad. The next day, 34 customers were hospitalized for botulism; two died. Before this incident, the Food and Drug Administration (FDA) defined Potentially Hazardous Foods (PHF) as any moist, high protein food capable of supporting rapid germ growth. After this deadly incident, the FDA included cooked potatoes in the list of PHFs!

▲ Just how did those people get botulism from potato salad?

Answers to the case studies can be found in Appendix A.

Discussion Questions (Short Answer)

1. Briefly explain the difference between an infection, an intoxication, and a toxin-mediated infection. Give an example of each.

2. What groups of people have a greater risk of acquiring foodborne illness?

3. Identify the three categories of foodborne illness hazards and give an example of each.

4. What type of biological hazard should be of greatest concern to food establishment managers? Why?

5. What does FATTOM represent?

6. At what pH and water activity (A_w) levels do bacteria grow best?

7. What is the food temperature danger zone?

8. What is a bacterial spore?

9. What is a potentially hazardous food? What three characteristics do they share?

10. What are viruses? What are parasites?

Quiz 2.1(Multiple Choice)

1. Bacteria are the most common cause of foodborne disease in a food establishment because:

   a. Under ideal conditions, they can grow very rapidly.

   b. Bacteria are found naturally in many foods.

   c. Bacteria can be easily transferred from one source to another.

   d. All of the above.

2. Most of the bacteria that cause foodborne illness grow:

   a. With or without oxygen at an ideal temperature of 98.6° F (39°C).

   b. Only without oxygen at an ideal temperature of 110°F (43°C).

   c. Only with oxygen at an ideal temperature of 110°F (43°C).

   d. Only without oxygen at an ideal temperature of 98.6°F (39°C).

3. Which of the following organisms are most likely to cause a foodborne disease outbreak in a food establishment?

   a. *Salmonella spp.* and *Campylobacter jejuni.*

   b. *Crytosporidium parvum* and *Giardia lamblia.*

   c. Hepatitis A and Norwalk virus group.

   d. *Anisakis simplex.* and *Trichinella spiralis.*

4. Bacteria grow best within a narrow temperature range called the temperature danger zone. The temperature danger zone is between:

   a. 0° and 220°F (-18°C and 104°C).

   b. 0° and 140°F (-18°C and 60°C).

   c. 41° and 140°F (5°C and 60°C).

   d. 41° and 220°F (5°C and 104°C).

5. Bacteria that cause foodborne illness will only grow on foods that have a pH at _____ or above and a water activity (A_w) at _____ or above.

    a. 3.2; .85
    b. 4.6; .85
    c. 6.5; .80
    d. 8.0; .70

6. Which of the following bacteria produce a toxin that is most likely to cause death if consumed?

    a. *Campylobacter jejuni.*
    b. *Clostridium botulinum.*
    c. *Escherichia coli* O157:H7.
    d. *Listeria monocytogenes.*

7. Some bacteria form spores to help them

    a. Reproduce.
    b. Move more easily from one location to another.
    c. Survive adverse environmental conditions.
    d. Grow in high acid foods.

8. Which of the following is a histamine poisoning?

    a. Ciguatoxin.
    b. Scombrotoxin.
    c. Mycotoxin.
    d. Paralytic Shellfish Poisoning (PSP).

9. Which of the following is **not** considered a potentially hazardous food group?

    a. Red meats.
    b. Fish and shellfish.
    c. Poultry and eggs.
    d. Dried grains and spices.

10. The most effective way to control the growth of bacteria in a food establishment is by controlling:

    a. Time and temperature.
    b. pH and oxygen conditions.
    c. Temperature and water activity.
    d. Time and food availability.

Answers to the multiple choice questions are provided in Appendix A.

References/Suggested Readings

Benenson, A. S. (Ed.) (1995). *Control Of Communicable Diseases in Man* (16th ed.). American Public Health Association. Washington, DC.

Cichy, R. F. (1994). *Quality Sanitation Management.* Educational Institute of the American Hotel and Motel Association. East Lansing, MI.

The Educational Foundation (1998). *ServSafe.* National Restaurant Association. Chicago, IL.

Food and Drug Administration (1999). *1999 Food Code.* United States Public Health Service. Washington, DC.

Kolchenar, L. H., and R. Donnelly (1994). *Quantity Food Sanitation.* Macmillan. New York, NY.

Longrèe, K., and G. Armbruster (1996). *Quantity Food Sanitation.* Macmillan. New York, NY.

Rue, N., National Assessment Institute (1994). *Handbook for Safe Food Service Management.* Prentice Hall. Englewood Cliffs, NJ.

Suggested Web Sites

Gateway to Government Food Safety Information
   www.foodsafety.gov

Centers for Disease Control and Prevention (CDC)
   www.cdc.gov

Bad Bug Book
   vm.cfsan.fda.gov/~mow/intro.html

USDA/FDA Food and Nutrition Information Center
   www.nal.usda.gov/fnic/

Disease Fact Sheets
   www.ph.dhr.stat.ga.us/epi/news/resources/fact.sheet.index.htm

The Food Allergy Network
   www.foodallergy.org

The National Food Safety Database
  www.foodsafety.org/index.htm

Food Safety at Iowa State University Extension
  www.extension.iastate.edu/foodsafety

United States Department of Agriculture (USDA)
  www.usda.gov

United States Food and Drug Administration (FDA)
  www.fda.gov

United States Environmental Protection Agency (EPA)
  www.epa.gov

Partnership for Food Safety Education
  www.fightbac.org

3 FACTORS THAT AFFECT FOODBORNE ILLNESS

Campylobacter jejuni *Outbreak Associated with Cross Contamination of Food*

Fourteen patrons in an Oklahoma town became ill with a Campylobacter jejuni infection after eating at a local eatery. The infected individuals experienced a variety of symptoms including diarrhea, fever, nausea, vomiting, abdominal cramps, and visible blood in their feces. All of the victims had eaten lettuce during their lunch, and 79% of them had eaten lasagna.

Inspection of the restaurant indicated that the countertop surface area in the kitchen was too small to separate raw poultry and other foods adequately during preparation. The cook reported cutting up raw chicken for the dinner meals before preparing salads, lasagna, and sandwiches as luncheon menu items. Lettuce for salads was shredded with a knife, and the cook wore a towel around her waist that she frequently used to dry her hands.

A chlorine bleach solution at the appropriate temperature and concentration was present to sanitize table surfaces, but it was uncertain whether the cook had actually cleaned and sanitized the countertop and utensils after cutting up the chicken.

Learning Objectives

After reading this chapter, you should be able to:

▲ Identify potential problems related to temperature abuse of foods

▲ Describe how to properly measure and maintain food temperatures to assure that foods are safe for consumption

▲ Identify potential problems related to a food worker's poor personal hygiene

▲ Explain how to improve personal hygiene habits to reduce the risk of foodborne illness

▲ Identify potential problems related to cross contamination of food

▲ Discuss procedures and methods to prevent cross contamination.

Essential Terms

Calibrate	Measuring device
Centers for Disease Control and Prevention (CDC)	Personal hygiene
Dial face bi-metal thermometer	Temperature abuse
Digital thermometer	Thermocouple
Food temperature	T-stick type melt devices
Impermeable	

Factors That Contribute to Foodborne Illness

The **Centers for Disease Control and Prevention (CDC)** is an agency of the Federal Government. One of CDC's primary responsibilities is to collect statistics about diseases that affect people in America. The list includes foodborne illness. CDC statistics show that most outbreaks of foodborne disease occur because *food is mishandled*. The leading factors that cause foodborne illness, as identified by the CDC, are presented in Figure 3.1.

Almost all foodborne illnesses are linked to:

▲ Time and temperature abuse

▲ Poor personal hygiene and improper handwashing

▲ Cross contamination

▲ Contaminated ready-to-eat foods such as salad items and processed meats.

It is extremely important that you recognize these as major contributors to food contamination. This chapter describes prevention strategies necessary for safe food management.

4% Use of leftovers

7% Improper cleaning

7% Cross contamination

11% Contaminated raw food

12% Inadequate reheating

16% Improper hot storage

16% Inadequate cooking

20% Infected persons touching food

21% Time between preparing and serving

40% Improper cooling of foods

Figure 3.1—Factors Leading to Foodborne Illness (ranked by percentage of outbreaks)

(Source: Surveillance for Foodborne Disease Outbreaks—United States, 1988-1992)

Note: The total of the percentages is more than 100% because more than one factor may contribute to a foodborne disease outbreak.

What Are Time and Temperature Abuse?

Controlling temperature is perhaps the most critical way to ensure food safety. Most cases of foodborne illness can in some way be linked to **temperature abuse.** The term temperature abuse is used to describe situations when foods are:

▲ Exposed to temperatures in the danger zone for enough time to allow growth of harmful microorganisms

▲ Not cooked or reheated sufficiently to destroy harmful microorganisms.

In Chapter 2, *Hazards to Food Safety,* you learned that harmful microbes can grow in potentially hazardous foods when temperatures are between 41° and 140° F (5° and 60°C).

An important rule to remember for avoiding temperature abuse is **Keep Hot Foods Hot**, **Keep Cold Foods Cold**, or **Don't Keep the Food at All**.

Keep food temperatures, the temperature inside the core of a food item, above the temperature danger zone (140°F, 60°C) to prevent harmful microbes from growing (Figure 3.2). Higher temperatures destroy microbes. However, toxins produced by microbes may or may not be affected by heat.

Keep food temperatures below the temperature danger zone (41°F, 5°C) to prevent most microbes from growing. Bacteria that can grow in lower temperatures, do so very slowly.

Figure 3.2—Temperature Danger Zone

There are unavoidable situations during food production when foods must pass through the temperature danger zone such as:

▲ Cooking

▲ Cooling

▲ Reheating

▲ Food preparation (slicing, mixing, etc.).

During these activities, you must minimize the amount of time foods are in the temperature danger zone to control microbial growth. When it is necessary for a food to pass through the temperature danger zone, do it as quickly as possible. In addition, foods should pass through the danger zone as few times as possible. As a rule, hot foods should be cooled and reheated only one time. If not completely used up after the first reheat the food should be discarded.

Cooking foods to make them more enjoyable to eat began with the discovery of fire. Heating improves texture and flavor and also destroys harmful microorganisms. As you learned in Chapter 2, many raw foods naturally contain harmful microbes or can become contaminated during handling. When you cook and reheat foods properly, microbes are reduced to safe levels or are destroyed. *Cooking and reheating are two very important processes for safe food management.*

How and When to Measure Food Temperatures

Maintaining safe food temperatures is an essential and effective part of food safety management. You must know *how* and *when* to measure food temperatures correctly to prevent temperature abuse. Thermometers, thermocouples, or other devices are used to measure the temperature of stored, cooked, hot-held, cold-held, and reheated foods.

Thermometers

The first item you will need is an approved **food temperature measuring device**. Four devices commonly used to measure food, water, and air temperatures are shown in Figure 3.3.

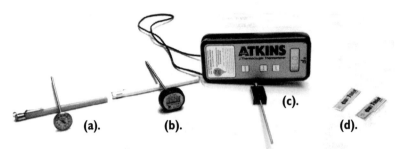

(a.) Dial-face thermometer (b.) Digital thermometer

(c). Thermocouple thermometer (d). T-sticks

Figure 3.3—Temperature Measuring Devices

The most common type of thermometer is the *dial-faced, metal stem-type (bi-metallic) thermometer*. [See Figure 3.3(a).] The metal-stemmed thermometer probe is at least five inches long with a one-inch diameter dial face. Use this type of thermometer to measure internal food temperatures at every stage of food preparation.

The *digital thermometer* shows the temperature numerically on a digital display. [See Figure 3.3(b).]

Thermocouple thermometers provide a digital readout of the temperature and have a variety of immersion, surface, and specialty probes that can be interchanged to suit most applications. [See Figure 3.3(c).]

The infrared thermometer is the newest type of temperature measuring device to be used in retail food establishments. Unlike bi-metallic metal stem and thermocouple thermometers, infrared thermometers measure the surface temperature of foods without actually coming into contact with the food. This virtually eliminates the possibility of cross-contamination. Infrared thermometers are valuable "screening" devices because they allow food industry professionals to measure the temperature of many different products in a very short period of time. The accuracy of an infrared thermometer should be checked frequently with a crushed ice and water solution. If the temperature measured by the thermometer is several degrees off from 32°F (0°C), the device should be sent back to the company for calibration. The lens of the thermometer should be cleaned at least monthly with a cotton swab and distilled water. The most significant limitation found with infrared thermometers is "thermal shock." In other words, you can not use an infrared thermometer to measure temperatures in a walk-in cooler or freezer and immediately start measuring temperatures of hot foods on a stove or hot-holding area. It takes about 20 minutes for the device to adjust to the temperature of the surroundings.

T-sticks are examples of *melt devices.* [See Figure 3.3(d).] This single-use disposable thermometer measures only one temperature. The wax melts when the temperature reaches or exceeds the set point and the color changes (usually from white to black). Melt devices may be used to monitor product temperatures and the sanitizing temperature in dishwashing machines.

Maximum registering (holding) thermometers measure the temperature of hot water used to sanitize dishware and utensils in mechanical warewashing machines.

Bi-metal thermometers measure temperatures ranging from 0° to 220°F (-18° to 104°C) with 2°F increments. Digital and thermocouple thermometers typically measure a much wider range of temperatures. Food temperature measuring devices scaled only in Celsius or dually scaled in Celsius and Fahrenheit must be accurate to +1.8°F (+1°C). Food temperature measuring devices scaled in Fahrenheit only must be accurate to +2°F.

Mercury filled and glass thermometers should not be used in food establishments. The mercury is very toxic and glass is a physical hazard if it gets into food and is consumed.

Clean and sanitize thermometers properly to avoid contaminating food that is being tested. This is very important *when testing raw* and *then ready-to-eat food items.* To clean and sanitize a food thermometer, wipe off any food particles, place the stem or probe in sanitizing solution for at least 5 seconds, then air dry. When monitoring *only raw* foods, or *only cooked* foods being held at 140°F (60°C), wipe the stem of the thermometer with an alcohol swab between measurements.

When and How to Calibrate Thermometers

Calibrate dial face metal stem-type (bi-metal) thermometers:

▲ Before their first use

▲ At regular intervals

▲ If dropped

▲ If used to measure extreme temperatures

▲ Whenever accuracy is in question.

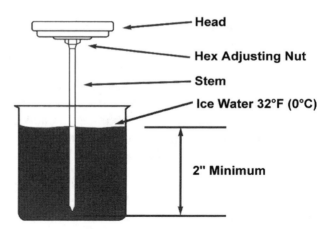

Figure 3.4—Calibrating a Dial-faced Thermometer
(Source: Cooper Instrument Corporation)

Most dial-faced thermometers have a *calibration nut* just behind the dial (Figure 3.4). **Calibrate** dial-faced thermometers by the boiling point or ice point method. See specific directions for calibration as described in the following section. Use pliers or an open-ended wrench to adjust the indicator needle.

Boiling Point Method

Immerse at least the first two inches of the stem from the tip (the sensing part of the probe) into boiling water, and adjust the needle to 212°F (100°C). At higher altitudes, the temperature of the boiling point will vary. Consult your local health department if you have any questions.

Ice Point Method

Insert the probe into a cup of crushed ice. Add enough cold water to remove any air pockets that might remain. Let the probe and ice mixture stabilize and adjust the needle to 32°F (0°C).

Measuring Food Temperature

The sensing portion of a food thermometer is at the end of the stem or probe. On the bi-metal thermometer, the sensing portion extends from the tip up to the "dimple" mark on the stem. An average of the temperature is measured over this distance. The sensing portion for digital and thermocouple thermometers is closer to the tip.

a. in a salad bar

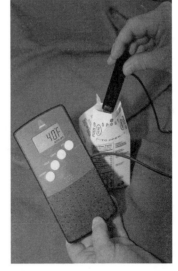

d. for single use containers

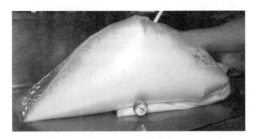

b. for bulk storage of milk

c. for pre-packaged meats

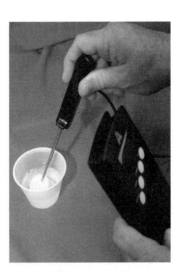

e. for raw shell eggs

Figure 3.5—Measuring Food Temperature

Accurate readings are only possible when the sensing portion of the temperature measuring device is inserted deeply into the food. For bi-metal thermometers, immerse the needle tip at least 2 inches into the material to measured. For the digital and thermocouple thermometers, the tip must be immersed 1" or more. Always insert the sensing element of the thermometer into the center or thickest part of the food. When possible, stir the food before measuring the temperature. *Always wait for the temperature reading to stabilize.* [See Figure 3.5(a).].

The approximate temperature of packaged foods can be measured accurately without opening the package. Place the stem or probe of the thermometer between two packages of food or fold the package around the stem or probe to make good contact with the packaging [Figures 3.5(b) and (c)]. For some prepackaged foods, it may be best to open the package or container and measure the temperature of the food [Figures 3.5(d) and 3.5(e)]. To accurately and safely measure food temperatures, be sure to:

▲ Use an approved temperature measuring device that measures temperature from 0° to 220°F(-18° to 104°C)

▲ Locate the sensing portion of the measuring device

▲ Calibrate the measuring device using the ice or boiling point method

▲ Clean and sanitize the probe of the temperature measuring device according to procedure

▲ Measure the internal temperature of the food by inserting the probe into the center or thickest part of the item.

Preventing Temperature Abuse

Controlling temperatures in potentially hazardous food is important in almost all stages of food handling. Thus, measuring temperatures of potentially hazardous food is an important duty for nearly all food workers. The following table (Figure 3.6) lists safe temperature guidelines with the rationale for each step. Each of these guidelines will be discussed throughout the remainder of this book.

Steps in the Flow of Food	Safe Temperature Guidelines	Rationale for Temperature Guidelines
Receiving and Storing Frozen Potentially Hazardous Foods	Foods should be frozen solidly and maintained frozen at all times.	Proper freezing of foods helps to maintain food quality and prevents the growth of spoilage and harmful microorganisms.
Receiving and Storing Refrigerated Potentially Hazardous Foods	Foods should be received and stored so that food is always at or below 41°F (5°C). Raw shell eggs may be received at 45°F (7°C) or below.	Receiving and storing foods below 41°F (5°C) prevents or slows the growth of harmful microorganisms.
Cooking Potentially Hazardous Foods	Different foods, and the methods by which they are cooked, require different end point temperatures to be safe. The range of safe cooking temperatures can vary from 145° to 165°F (63° to 74°C). Foods should reach the required final cooking temperature within 2 hours.	Proper cooking destroys harmful microorganisms that may be present in the food.
Cooling Potentially Hazardous Foods	During cooling, food must be cooled from 140° to 70°F (60° to 21°C) within 2 hours and from 70° to 41°F (21° to 5°C) within an additional 4 hours.	Proper cooling prevents the conversion of sporeforming bacterial cells to vegetative bacterial cells, and the growth of vegetative bacterial cells.
Reheating Potentially Hazardous Foods	All reheated foods must be reheated to at least 165°F (74°C) within 2 hours.	Proper reheating destroys harmful bacteria that may be present in foods.

Figure 3.6—Safe Temperature Guidelines for Various Steps in the Flow of Food

Steps in the Flow of Food	Safe Temperature Guidelines	Rationale for Temperature Guidelines
Hot-holding Potentially Hazardous Foods	All foods must be cooked to a safe temperature and then held at greater than 140°F (60°C).	Proper holding of food prevents the growth of harmful bacteria.
Cold-Holding Potentially Hazardous Foods	All foods that are held and served cold must be held at 41°F (5°C) or below.	Holding cold foods below 41°F (5°C) prevents or slows the growth of harmful microorganisms.
Thawing Potentially Hazardous Foods	Thawing may be done in a refrigerator at 41°F (5°C) or less, in a microwave oven & then immediately cooked, or under cold running water 70°F(21°C) in two hours or less.	Proper thawing prevents or reduces the growth or harmful bacteria.
Food Preparation	During food preparation, food should only be in the temperature danger zone (between 41° and 140° F, 5° and 60°C) for a maximum total time of 4 hours.	Maintaining foods between 41° and 140°F (5° and 60°C) for no more than 4 hours limits the number of microorganisms that can grow.

Figure 3.6—Safe Temperature Guidelines for Various Steps in the Flow of Food (Cont.)

Keep Cold Foods Cold and Hot Foods Hot!

Frozen foods should be kept solidly frozen until they are ready to be used. Freezing helps to retain product quality. Proper frozen food temperatures do not permit disease-causing and spoilage microorganisms to grow. Cold temperatures also help to preserve the color and flavor characteristics. Frozen foods can be stored for long periods of time without losing their wholesomeness and quality.

Refrigerated foods are held cold, not frozen. Cold foods should be maintained at 41°F (5°C) or below. Do not forget that some harmful bacteria and many spoilage bacteria can grow at temperatures below 41°F (5°C), although their growth is very slow (Longrèe and Armbruster, 1996). By keeping cold foods at 41°F (5°C) or below, you can reduce the growth of most harmful microorganisms and extend the shelf-life of the product. For maximum quality and freshness, hold cold foods for the shortest amount of time possible.

Applying heat is another method used to preserve food. To destroy harmful bacteria, heat food to proper temperatures. Established safe cooking temperatures are based on the type of food and the method used to heat the product. Cooked foods that have been cooled and then reheated must be maintained at 140°F (60°C) or above until used. You must keep foods hot to stop growth of harmful bacteria.

There are times during food production when foods must be in the temperature danger zone. Recognize that the time spent in the temperature danger zone should be minimal for potentially hazardous items (Figure 3.7).

Improper cooling of foods is the number one contributing factor that leads to foodborne illness. Spores of certain bacteria like *Clostridium botulinum, Clostridium perfringens*, and *Bacillus cereus* can survive cooking temperatures. Remember, if spores survive and are exposed to ideal conditions, they can again become vegetative cells (Figure 2.2).

The *Food Code* contains cooling guidelines that permit foods to be in the temperature danger zone for a total of 6 hours. The *Food Code* specifically states that foods must be cooled from 140° to 70°F (60° to 21°C) in 2 hours, and 70° to 41°F (21° to 5°C) in an additional 4 hours.

The industry standard for cooling requires food temperatures to go from 140°F (60°C) to 40°F (5°C) in 4 hours.

To destroy many of the bacteria that may have grown during the cooling process, reheat foods to 165°F (74°C) within two hours to

prevent the number of organisms from reaching levels that can cause foodborne illness.

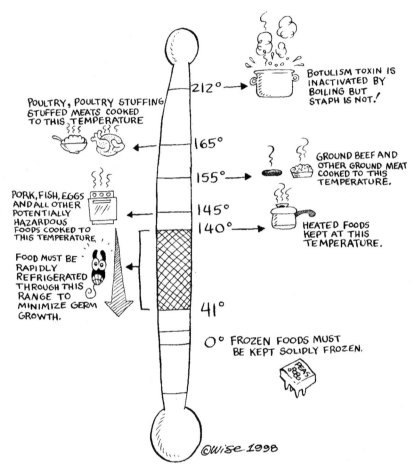

POULTRY, POULTRY STUFFING STUFFED MEATS COOKED TO THIS TEMPERATURE

212° → BOTULISM TOXIN IS INACTIVATED BY BOILING BUT STAPH IS NOT.!

165°

155° → GROUND BEEF AND OTHER GROUND MEAT COOKED TO THIS TEMPERATURE.

PORK, FISH, EGGS AND ALL OTHER POTENTIALLY HAZARDOUS FOODS COOKED TO THIS TEMPERATURE.

145°

140° → HEATED FOODS KEPT AT THIS TEMPERATURE.

FOOD MUST BE RAPIDLY REFRIGERATED THROUGH THIS RANGE TO MINIMIZE GERM GROWTH.

41°

0° FROZEN FOODS MUST BE KEPT SOLIDLY FROZEN.

©Wise 1998

Figure 3.7—Keep it Hot, Keep it Cold, or Don't Keep It

The preferred method for thawing foods is in the refrigerator at 41°F (5°C) or below. This prevents the food from entering the food temperature danger zone.

Other acceptable methods for thawing include using a microwave oven, as a part of the cooking process, or place food under cold running water. Proper thawing reduces the chances for bacterial growth, especially on the outer surfaces of food.

More detailed strategies for minimizing the amount of time a food is in the temperature danger zone during cooling, thawing, and food preparation will be presented in the next chapter.

The Importance of Handwashing and Good Personal Hygiene

The cleanliness and personal hygiene of food workers are extremely important. If a food worker is not clean, the food can become **contaminated**.

As you know, even healthy humans can be a source of harmful microorganisms. Therefore, **good personal hygiene is essential for those who handle foods. Desirable behaviors include:**

▲ Knowing when and how to properly wash hands

▲ Wearing clean clothing

▲ Maintaining good personal habits (bathing; washing and restraining hair; keeping fingernails short and clean; washing hands after using toilet; etc.)

▲ Maintaining good health and reporting when sick to avoid spreading possible infections.

Just think of all the things your hands touch during a typical work day. You may take out the trash, cover a sneeze, scratch an itch, or mop up a spill. When you touch your face or skin, run your fingers through your hair or a beard, use the toilet, or blow your nose, you transfer potentially harmful germs to your hands. *Staphylococcus aureus*, Hepatitis A, and *Shigella* spp. are examples of pathogens that may be found in and on the human body and can be transferred to foods by hand contact.

Personnel involved in food preparation and service must know how and when to wash their hands. Figure 3.8 provides a list of situations when handwashing is required.

Using approved cleaning compounds (soap or detergent), vigorously rub surfaces of fingers and fingertips, front and back of hands, wrists, and forearms for at least 20 seconds. Remember, soap, warm water, and friction are needed to adequately clean skin. A significant number of germs are removed by friction alone. A brush can be helpful when cleaning hands. However, the brush must be kept clean and sanitary.

Always wash hands:

▲ Before food preparation

▲ After touching human body parts

▲ After using the toilet

▲ After coughing, sneezing, using a handkerchief or tissue, using tobacco, eating, or drinking

▲ During food preparation when switching between raw foods and ready-to-eat products

▲ After engaging in any activities that may contaminate hands (taking out the garbage, wiping counters or tables, handling cleaning chemicals, picking up dropped items, etc.)

▲ After caring for or touching animals.

Figure 3.8—Handwashing Guidelines
(Source: *Food Code*)

Thoroughly rinse under clean warm running water; and clean under and around fingernails and between fingers.

Dry hands using a single service paper towel, an electric hand dryer, or clean section of continuous rolled cloth towel (if allowed in your area). *Do not dry hands on your apron or a dish towel.* (See the handwashing procedure in Figure 3.9.)

In addition to proper handwashing, fingernails should be trimmed, filed, and maintained so that handwashing will effectively remove soil from under and around them. Unless wearing intact gloves in good repair, a food employee must not wear fingernail polish or artificial fingernails when working with exposed food.

According to the *Food Code*, hands shall be washed in a separate sink specified as a handwashing sink (Figure 3.10). An automatic handwashing facility may be used by food workers to clean their hands. However, the system must be capable of removing the types of soils encountered in the food operation.

Each handwashing sink must be provided with hand cleanser (soap or detergent) in a dispenser and a suitable hand-drying device.

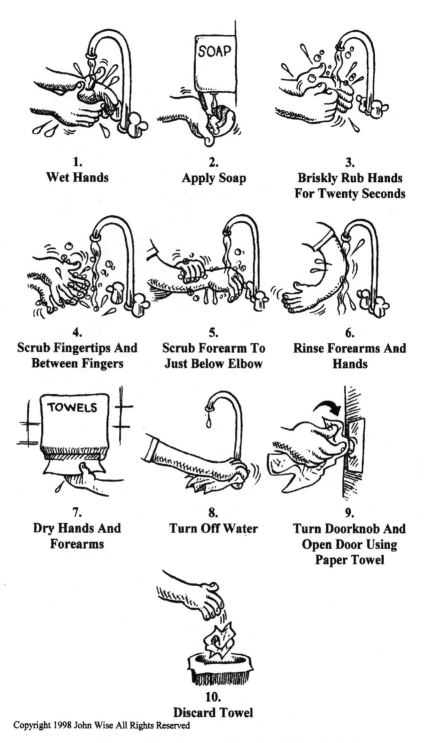

1.
Wet Hands

2.
Apply Soap

3.
Briskly Rub Hands
For Twenty Seconds

4.
Scrub Fingertips And
Between Fingers

5.
Scrub Forearm To
Just Below Elbow

6.
Rinse Forearms And
Hands

7.
Dry Hands And
Forearms

8.
Turn Off Water

9.
Turn Doorknob And
Open Door Using
Paper Towel

10.
Discard Towel

Figure 3.9—Proper Handwashing Technique

Figure 3.10—Handwashing Station

 Hand sanitizing lotions and chemical hand sanitizing solutions may be used by food employees *in addition to handwashing*. Proper washing helps to remove visible hand dirt and the microorganisms it contains. **Hand sanitizing lotions must never be used as a replacement for handwashing**.

After washing, the sanitizing chemical remains on the hands after application because the hands do not receive additional washing or rinsing. It is critical that hand sanitizers be formulated with safe ingredients because it is likely that a food employee's hands will touch food, food contact surfaces, or equipment and utensils after using the product.

Except when washing fruits and vegetables, food workers should not contact exposed ready-to-eat foods with their hands. Instead, they should use suitable utensils such as deli-tissue, spatulas, and tongs. Also, minimize bare hand and arm contact with exposed food that is not in ready-to-eat form.

Food establishments sometimes allow their food workers to use disposable gloves to help prevent contamination of foods. Gloves protect food from direct contact by human hands. Gloves must be impermeable, therefore not allowing anything to penetrate the porous texture of the glove. **You must treat disposable gloves as a second skin. Whatever can contaminate a human hand can also contaminate a disposable glove. Therefore, whenever hands should be washed, a new pair of disposable gloves should be worn (Figure 3.11).**

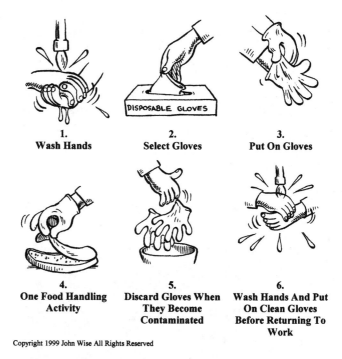

1.
Wash Hands

2.
Select Gloves

3.
Put On Gloves

4.
One Food Handling Activity

5.
Discard Gloves When They Become Contaminated

6.
Wash Hands And Put On Clean Gloves Before Returning To Work

Figure 3.11—Proper Use of Disposable Gloves

For example, if you are wearing disposable gloves and handling raw food, you must discard those gloves, wash your hands, and put on a fresh pair of gloves before you handle ready-to-eat foods.

Never handle money with gloved hands unless you immediately remove and discard the gloves. Money is highly contaminated from handling. If you take latex gloves off by rolling them inside out, the inner surface of the glove is very contaminated from your skin. Again, if you take disposable gloves off, throw them away. **Never reuse or wash disposable gloves—always throw them away.**

Outer Clothing and Apparel

Work clothes and other apparel should always be clean. The appearance of a clean uniform is more appealing to your guests.

During food preparation and service, it is easy for a food worker's clothing to become contaminated. If you feel that you have contaminated your outer clothing, change into a new set of work clothes.

For example, if you normally wear an apron and work with raw foods, put on a fresh, clean apron before working with ready-to-eat foods. Also, never dry or wipe your hands on the apron. As soon as you do that, the apron is contaminated. Smocks and aprons help to reduce transfer of microbes to exposed food.

Keep in mind, however, that **protective apparel is similar to a disposable glove. They no longer protect food when contaminated.**

Hats, hair coverings or nets, and beard restraints discourage workers from touching their hair or beard. These restraints also prevent hair from falling into food or onto food-contact surfaces. Most state and local jurisdictions require food workers to effectively keep their hair from contacting exposed food, clean equipment, utensils and linens, and unwrapped single-use articles.

Figure 3.12—Employees Must Wear Proper Attire during Food Preparation and Service

Personal Habits

Personal hygiene means good health habits including bathing, washing hair, wearing clean clothing, and frequent hand washing. Poor personal habits are serious hazards in food establishments (Figure 3.13). A food worker's fingers may be contaminated with saliva during eating and smoking. Saliva, sweat, and other body fluids can be harmful sources of contamination if they get into food.

Supervisors should enforce rules against eating, chewing gum, and smoking in food preparation, service, and warewashing areas. The *Food Code* permits food workers to drink water to prevent dehydration. The water must be in a covered container with a straw. The container must be handled in a way that prevents contamination of the employee's hands, the container, exposed food, equipment, and single-use articles.

Jewelry has no place in food production and warewashing areas. Rings, bracelets, necklaces, earrings, watches, and other body part ornaments can harbor germs that can cause foodborne illness. Jewelry can also fall into food causing a physical hazard. A plain wedding band is the only piece of jewelry that may be worn in food production and warewashing areas.

Figure 3.13—Don't Set a Bad Example

Personal Health

In an attempt to reduce the risk caused by sick food workers (Figure 3.14), **the *Food Code* requires employees to report to the person in charge when they have been diagnosed with:**

▲ *Salmonella typhi*

▲ *Shigella spp.*

▲ *E. coli* O157:H7

▲ Hepatitis A

▲ Symptoms of intestinal illness or flu-like symptoms (such as vomiting, diarrhea, fever, sore throat, coughing, sneezing, running nose, or jaundice).

If a food worker is directly or indirectly exposed to *Salmonella typhi, Shigella spp., E. coli* O157:H7, or Hepatitis A, it must be reported to the supervisor. All of these diseases are easily transferred to foods and are considered severe health hazards. The person in charge shall notify the regulatory authority that a food employee is diagnosed with an illness due to *Salmonella typhi, Shigella spp., E. coli* O157:H7, or Hepatitis A virus.

Figure 3.14—Sick Employees Must Not Work With Food

If you have been exposed to any of these agents, you may be excluded from work or be assigned to restricted activities having no food contact. Employees and workers diagnosed with one of these diseases must not handle exposed food or have contact with clean equipment, utensils, linens, or unwrapped single-service utensils.

Staphylococcus aureus bacteria are often found in infected wounds, cuts, and pimples. Infected wounds should be completely covered by a dry, tight-fitting, impermeable bandage. Cuts or burns on a food worker's hands must be thoroughly bandaged and covered with a clean disposable glove.

When a foodborne illness outbreak occurs and an employee is suspected of carrying the disease, he or she must not work until certified as safe by a licensed physician, nurse practitioner, or physician assistant. Exclusions and restrictions for those exposed to or experiencing symptoms of a foodborne illness are specified in the *Food Code*. **The health of food workers is extremely important in safe food management.**

To date, there has not been a medically documented case of Acquired Immune Deficiency Syndrome (AIDS) transmitted by food. Therefore, AIDS is not considered a foodborne illness.

The Americans with Disabilities Act (ADA) prohibits discrimination against people with disabilities in jobs and public accommodations. Employers may **not** fire or transfer individuals who have AIDS or test positive for the HIV virus away from food handling activities. Employers must also maintain the confidentiality of employees who have AIDS or any other illness.

Cross Contamination

Contaminated food contains germs or harmful substances that can cause foodborne illness. The transfer of germs from one food item to another is called cross contamination. This commonly happens when germs from raw food are transferred to a cooked or ready-to-eat food via contaminated hands, equipment, or utensils.

For example, bacteria from raw chicken can be transferred to a ready-to-eat food such as lettuce or tomato when the same cutting board is used without being washed and sanitized between foods (Figure 3.15).

Figure 3.15—Cross Contamination of a Cutting Board

Cross contamination also happens when raw foods are stored above ready-to-eat foods. Juices from the raw product can drip or splash onto a ready-to-eat food as in Figure 3.16.

In a food establishment, germs can be transferred by a food worker, equipment and utensils, or another food. Remember to keep things **clean**—removal of the soil—and **sanitary** in order to reduce the disease causing microorganisms to a safe level. Therefore, it is extremely important to:

▲ Always store cooked and ready-to-eat foods over raw products

▲ Keep raw and ready-to-eat foods separate during storage

▲ Use good personal hygiene and handwashing

▲ Keep all food contact surfaces clean and sanitary.

Avoiding Cross Contamination

There are many things you can do to avoid, or at least reduce, opportunities for cross contamination during food preparation and service. An important part of food safety is to impress upon your employees the importance of clean hands and clothing. *Stress the importance of when and how to properly wash hands.* Provide ongoing supervision to ensure that employees remain clean and apply safe food handling practices when working with food.

To avoid cross contamination by way of food contact surfaces (i.e., cutting boards, knife blades, slicers, and preparation tables) an effective in-house cleaning and sanitation program is important. You will learn more on this subject in Chapter 7, Cleaning and Sanitizing Operation*s*.

**Figure 3.16—Never Put Raw Food Over Cooked
or Ready-to-Eat Food**

If a food contact surface is not properly cleaned and sanitized, bacteria can accumulate and grow on the surface. Once bacteria get on the surface, they can be difficult to remove. To avoid such buildup, *always* clean and sanitize food contact surfaces thoroughly before working with ready-to-eat foods and foods that will not be heat processed.

Preventive measures eliminate the possibility of cross contamination between products and may include the following:

▲ Use of separate equipment, such as cutting boards, for raw foods and ready-to-eat foods (color coding may be helpful for this task)

▲ Use of clean, sanitized equipment and utensils for food production

▲ Preparation of ready-to-eat foods first—then raw foods

▲ Preparation of raw and ready-to-eat foods in separate areas of the kitchen.

Always keep raw foods separate from ready-to-eat foods. In the refrigerator, ready-to-eat foods must be stored above raw foods. Display cases, such as those used to display seafood items, should be designed to keep raw and cooked food items separate. In addition, separate buckets inplace of sanitizing solution and wiping cloths should be used for cleaning food contact surfaces in raw and ready-to-eat food production areas.

Other Sources of Contamination

Raw foods, such as fruits and vegetables, should be treated like ready-to-eat foods. Always wash these foods before use. Washing removes soil and other contaminants. Chemicals may be used to wash or peel raw whole fruits and vegetables. These chemicals must be non-toxic and meet the requirements set forth in the Code of Federal Regulations under title 21CFR173.315.

Animals are not allowed in food establishments unless they are being used for support or special service (i.e., guide dogs for the blind). It is very important that you do not touch animals during food preparation and service [Figure 3.17(a)]. If you should touch an animal for any reason, wash your hands before returning to work.

Germs from a worker's mouth can be transferred to food when the employee uses improper tasting techniques [Figure 3.17(b)]. A food worker may not use a utensil more than once to taste food that is to be sold or served.

Animals, rodents, and pests are common sources for food contamination. Rodents and pests usually enter food establishments during delivery or when garbage facilities are not properly maintained. A good integrated pest management (IPM) program should be established and maintained in any food establishment. You will learn about IPM programs in Chapter 8, Environmental Sanitation and Maintenance.

a. Touching Animals **b. Improper Tasting**

Figure 3.17—Other Sources of Contamination

Summary

The case study presented at the beginning of this chapter shows how bacteria and other germs can be transferred from raw to cooked or ready-to-eat foods via cross contamination. What conditions helped cause this outbreak? What controls should be implemented to prevent this situation?

Raw chicken is a common source of *Campylobacter* bacteria. An opportunity for cross contamination to occur in food establishments occurs when raw and ready-to-eat foods are frequently handled in the same food production areas. Failure to properly wash hands and clean and sanitize utensils and countertops can lead to the contamination of cooked and ready-to-eat foods with microbes from raw foods.

Almost all foodborne illness can be prevented with proper food handling practices. Most foodborne illnesses are due to temperature abuse, poor personal hygiene, handwashing practices, and cross contamination.

Temperature abuse can occur during receiving, storing, cooking, cooling, reheating, hot-holding, and cold-holding of foods. Depending on the food type, there are specific temperature requirements to ensure food safety. These requirements are discussed in the next chapter. **A good rule to follow in any food establishment is, "Keep It Clean! Keep It Hot! Keep It Cold! Or Don't Keep It!"**

Food workers should always use good health and hygiene practices. Clean clothing, hair restraints, and proper handwashing practices are fundamental in safe food management. Effective supervision and enforcement of proper procedures form the foundation of a successful food operation.

Control cross contamination with proper cleaning and sanitizing. Avoid cross contamination from one food to another. Keep foods separate and store raw foods below cooked and ready-to-eat foods. Always use proper food handling techniques.

Keep foods at proper temperature, use good personal hygiene, and control cross contamination. These are the essentials of safe food management.

Case Study 3.1

Michelle is in charge of all the cold salad preparation at a local hotel restaurant. Her main responsibility is to prepare green salads for the evening meal. When she arrived at work one day, the food manager on duty noticed that she was coughing and sneezing frequently. Michelle also indicated that she was battling a case of the flu.

▲ How should the food manager handle this situation?

Case Study 3.2

Cameron just purchased a new food thermometer. He checked to see that it was approved for use in food and it could measure 0° to 220° F (-18° to 104°C). He found the sensing area and then calibrated it properly. He was excited by his new purchase and wanted to see how well the thermometer performed. He first measured raw meats in the cold room. They all checked out well and measured between 35° and 41° F (2° and 5°C). As he exited the cold room, he immediately went to the customer self-service bar and measured the temperature of the cooked scrambled eggs. They measured 145° F (63°C). Cameron was satisfied because the cold foods were less than 41°F (5°C) and the hot foods measured higher than 140°F (60°C).

▲ What error did Cameron make when measuring food temperature?

Case Study 3.3

An outbreak of foodborne Hepatitis A occurred in a small Missouri town. One hundred and thirty people became ill and four died as a result of their illness. Cases stemming from this outbreak were reported as far away as Oklahoma, Florida, Alabama, and Maine.

A foodborne disease investigation revealed that all of the victims had consumed lettuce, either in salad or as a garnish. Investigators ruled out the theory that the lettuce had been contaminated at the source, because other restaurants supplied by the same source experienced no illnesses. In the end, the Missouri Department of Health Officials concluded that direct contamination of food by infected workers was the most likely cause of the outbreak. The initial source of the illness was believed to be a waiter who had been infected by his child in daycare. The waiter and four other food workers who tested positive for the Hepatitis A virus handled lettuce and were involved in preparing and serving salads.

▲ 1. What went wrong?

Answers to the case studies are provided in Appendix A.

Discussion Questions (Short Answer)

1. When purchasing and using a food thermometer, what are the six important things that you need to remember?

2. How frequently should thermometers be calibrated? Describe the two methods for calibrating food thermometers.

3. Under what conditions are foods "temperature abused"?

4. What is meant by *poor personal hygiene* and how can this lead to foodborne illness?

5. What does the term *cross contamination* mean?

Quiz 3.1 (Multiple Choice)

1. The number one contributing factor leading to foodborne illness in food establishments is:

    a. Improper cooling of foods.

    b. Cross contamination.

    c. Poor personal hygiene.

    d. Inadequate cooking of foods.

2. Foodborne illness can be caused by:

    a. Poor personal hygiene.

    b. Cross-contamination.

    c. Temperature abuse.

    d. All of the above.

3. Regarding food thermometers, which statement is **false**? They should

    a. Be calibrated.

    b. Measure temperatures between 41° and 140°F (5° and 60°C).

    c. Measure temperatures between 0°and 220°F (-18° and 104°C).

    d. Be approved for use in foods.

4. Good personal hygiene includes

    a. Using hand sanitizers instead of handwashing.

    b. Keeping hands and clothes clean and sanitary.

    c. Wearing attractive uniforms.

    d. Cleaning and sanitizing food contact surfaces.

5. Food workers should wash their hands after which of the following?

    a. Taking out the trash.

    b. Touching their face.

    c. Handling raw food.

    d. All of the above.

6. Cross contamination is a term used to describe the transfer of a foodborne hazard from one food to another:

    a. By a food worker's hands.

    b. From a cutting board.

    c. From a knife blade.

    d. All of the above.

7. A good way to prevent cross contamination of foods is to:

    a. Keep raw and cooked foods separate.

    b. Properly clean and sanitize food contact surfaces.

    c. Properly wash hands.

    d. All of the above.

8. Which of the following is an accepted personal hygiene practice?

    a. Wearing jewelry and false fingernails.

    b. Smoking and eating in food production areas.

    c. Wearing caps and hats.

    d. Wiping hands on a soiled apron.

9. After proper cooking, all foods that are to be held **cold** must be:

    a. Cooled quickly and held at 41°F (5°C) or below.

    b. Cooled quickly and held at 70°F (21°C) or above.

    c. Stored at room temperature until served.

    d. Cooled slowly and held at 50°F (10°C) or below.

10. After proper cooking, all foods that are to be held **hot** must be held at:

    a. 165°F (74°C) or above.

    b. 140°F (60°C) or above.

    c. Room temperature until served.

    d. 120°F (49°C) or above.

Answers to the multiple choice questions are provided in Appendix A.

References/Suggested Readings

Centers for Disease Control and Prevention (1996). *Surveillance for Foodborne Disease Outbreaks—United States, 1988-1992.* Atlanta, GA; U.S. Department of Health and Human Services.

Council for Agricultural Sciences and Technology (1995). Prevention of Food-borne Illness. Dairy. *Food and Environmental Sanitation.* Vol. 15(6), 357-367.

Food and Drug Administration (1999). *1999 Food Code.* United States Public Health Service. Washington, DC.

Food Safety and Inspection Service (1996). *Nationwide Broiler Chicken Microbiologic Baseline Data Collection Program, 1994-1995.* United States Department of Agriculture. Washington, DC.

Hecht, A. (1991). Preventing Foodborne Illness. *FDA Consumer.* Jan./Feb. (pp. 19-25).

Suggested Web Sites

Gateway to Government Food Safety Information
   www.foodsafety.gov

United States Department of Agriculture (USDA)
   www.usda.gov

Centers for Disease Control and Prevention (CDC)
   www. cdc.gov

United States Food and Drug Administration (FDA)
   www.fda.gov

USDA/FDA Food and Nutrition Information Center
   www.nal.usda.gov/fnic/

Partnership for Food Safety Education
   www.fightbac.org

Food Safety Day
   www.foodsci.purdue.edu/publications/foodsafetyday

4 FOLLOWING THE FOOD PRODUCT FLOW

Coleslaw linked to E. Coli O157:H7 *Outbreak*

Twenty-seven people suffered from E. coli O157:H7 when they ate con-taminated coleslaw at a food establishment near Indianapolis, Indiana. According to a health department report, employees at the establish-ment made a 100-pound batch of coleslaw using cabbage that was soft, heavily soiled, and had rotten leaves. The food establishment's standard operating procedure was for workers to remove the outer leaves from the heads of cabbage and then wash the heads before processing. However, workers at the establishment reported that the heads of cabbage that were used in the suspect batch of coleslaw had not been washed.

The unwashed heads of cabbage were cut into four pieces. The cabbage, carrots, and onions for the coleslaw were shredded and placed in a sani-tized plastic tub. The shredded vegetables were mixed with pre-pack-aged commercial dressing. Workers mixed the coleslaw by hand, wearing elbow-length disposable gloves.

The coleslaw was dispensed into 6 and 16-ounce containers. It was also sold on the lunch buffet table. Leftover coleslaw was stored in the plastic tub in the walk-in cooler. It was used as needed. The batch of coleslaw implicated in this outbreak was used for two and one-half days. Leftover coleslaw was not mixed with fresh portions.

Learning Objectives

After reading this chapter, you should be able to:

▲ Recognize codes and symbols used to designate food products that have been inspected by governmental agencies

▲ Apply purchasing and receiving procedures that enhance the protection of food products

▲ Evaluate equipment used to transport food products to food establishments

▲ Use approved devices to measure temperatures in food products safely and accurately

▲ Recognize product defects and refuse acceptance of products that do not meet established food safety criteria

▲ Identify product temperatures required at receiving and storage

▲ Discuss safe methods to thaw frozen foods

▲ Identify internal temperature requirements for cooking foods

▲ Explain the proper methods used to cool foods

▲ Discuss the importance of employee health and hygiene related to food flow

▲ Employ measures to prevent contamination and cross contamination of foods.

Essential Terms

Additives	Inspection for wholesomeness
Aseptic packaging	Irradiation
Cold-holding	Leftovers
Cooking	Modified Atmosphere Packaging (MAP)
Cooling	Pasteurization
First in, first out (FIFO)	Reheating
Grade standards	Sensory evaluation
Hermetic packaging	Sous-vide
Hot-holding	Wholesome

A Sound Food Supply

Effective purchasing paves the way for a successful food service operation. Purchasing is a highly skilled activity requiring knowledge of products and market conditions.

The main objectives of an effective purchasing program are to:

▲ Buy the product that is best suited for the job

▲ Buy the proper quantity of the item

▲ Pay the right price for the item

▲ Deal with only reputable and dependable suppliers.

Purchasing techniques include comparative shopping, evaluation of new products, wise judgment in timing large purchases of seasonal items, and selection of the most efficient supplier. In addition, a buyer must understand foods, specifications, formulations, and be able to evaluate these in terms of price and quality.

Purchase specifications are important to both buyer and management. They are the guidelines that detail the characteristics of a product, including such properties as:

▲ Quality grade

▲ Weight

▲ Count

▲ Contents

▲ Packaging.

Specifications make the task of comparison shopping easier, since the characteristics of a product are expressed in a common language and can be used as a basis for evaluation.

Buying from Approved Sources

You should always purchase food that is safe and **wholesome.** Something is wholesome if it is favorable to or promotes health. Food must come from an approved source that complies with all applicable local, state, and federal food laws.

Reputable suppliers deliver food products in vehicles that are clean and in good repair. These vehicles will also keep perishable foods at safe temperatures during transport. Reputable suppliers keep food products separate from general supplies (such as sanitizers or cleaning agents) during shipping. They also protect food packages from becoming damaged or torn. Receiving clerks should be instructed to inspect delivery vehicles. During these inspections, the clerk should check for:

▲ Cleanliness of the cargo area in the delivery vehicle

▲ Temperature of refrigerated and frozen storage areas (if applicable)

▲ Proper separation of food and nonfood items

▲ Signs of insect, rodent, or bird infestation.

Foods prepared in a private home are not considered to be from an approved source. They must not be used or offered for sale in a retail food establishment. Aunt Lou may make the best brownies in town, but unless the health department inspects and approves her kitchen, her homemade products may not be used, or offered for sale, in a food establishment. The use of "home-canned" food is prohibited because of a higher risk for foodborne illness.

In the United States, government agencies closely monitor and regulate food supplies to protect the public from foodborne illness. These agencies at the federal, state, and local levels are concerned with the protection of the food supply available to consumers. The fundamental concern of all these agencies is the safety and wholesomeness of the food supply as it reaches the customers. These agencies work in close cooperation with each other and the food industry. Additional information about these governmental agencies will be provided in Chapter 11, Food Safety Regulations.

Strategies for Determining Food Quality

Sensory evaluation is a commonly used method for making routine quality determinations on foods received at retail food establishments. This type of evaluation involves using the senses of smell, touch, sight, and sometimes taste.

As a first step, foods should be observed for color, texture, and visual evidence of spoilage. Quite often, spoilage is easily seen, as slime formation, mold growth, and discoloration. Product containers should also be checked for tears, punctures, dents, or other signs of damage.

Flavor is a combination of smell and taste. Spoiled foods frequently give off foul odors indicative of compounds such as ammonia and hydrogen sulfide (the smell of rotten eggs). These odors, caused by the breakdown of proteins through bacterial action, are usually very easy to smell.

The flavor of spoiled foods can range from loss of good characteristic taste to the development of objectionable tastes. Spoilage due to yeasts produces bubbles and an alcoholic flavor or smell. Milk develops an acidic taste and is often bitter when it spoils.

The feel of spoiled foods is frequently determined by the type of food and the spoilage organism involved. Some foods feel slimy, whereas others may feel mushy.

The quality and safety of a food are affected by many factors. A food that shows no signs of spoilage may not always be safe. Spoilage cannot be used as an indicator of food safety.

Measuring Temperatures at Receiving and Storage

Maintaining safe product temperature is a critical part of your food safety system. According to foodborne disease investigations, improper temperature control during food preparation and service is one of the leading contributors to foodborne illness. Temperature measuring devices are used in food establishments to measure temperatures of food, water, and the air of food storage areas (refrigerators, ovens, etc.).

You can measure the approximate temperature of packaged foods without opening the package. Place the probe of the thermometer or thermocouple between two packages of food or fold the package around the stem or probe to make good contact with the packaging. The photographs in Chapter 3, Figure 3.5, illustrate ways to measure the temperature of packaged and prepared foods.

To measure the temperature of an unpackaged food, insert the sensing element of the thermometer or thermocouple into the thickest part of the food. To avoid cross contamination of food items, be sure to wash, sanitize, and air dry the probe of the thermometer before going from a raw food to a cooked food.

Cold- or hot-holding equipment used for potentially hazardous foods must be equipped with an indicating or recording thermometer to measure the temperature of the storage environment. Equipment thermometers are either built into a piece of equipment or are fastened onto shelving or other apparatus. Position them where they can be easily read. **Place the sensor**

portion of the thermometer in the warmest part of a refrigeration
unit or in the coolest part of a hot food storage unit.

Following the Flow of Food

The flow of food onsite begins with receiving and storage. From
storage, food products move into the preparation and service
phases, which can involve one or more steps. A simple diagram
showing the flow of food is presented in Figure 4.1.

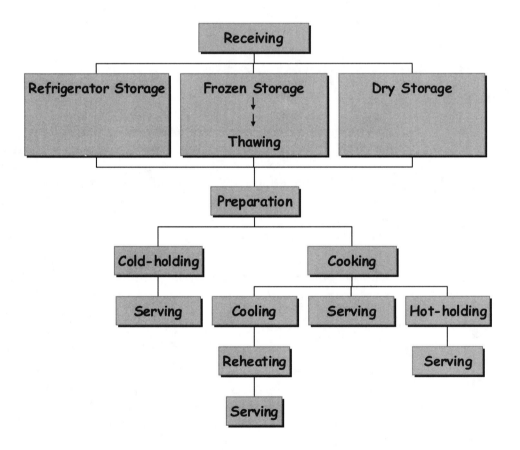

Figure 4.1—Food Flow Diagram

Receiving

Employees responsible for receiving products must carefully inspect
all incoming food supplies to make sure they are in sound

condition, free from filth or spoilage, and are at the proper temperatures.

As a first step, observe foods for color, texture, and visual evidence of spoilage. Quite often, spoilage is easily visible as in the case of slime formation, mold growth, and discoloration. Check the product containers for tears, punctures, dents, or other signs of damage.

Proper receiving requires a knowledgeable person who follows specific guidelines. Poor receiving procedures increase the chance of:

▲ Theft

▲ Acceptance of underweight merchandise

▲ Contamination

▲ Waste

▲ Acceptance of products that do not meet specifications.

Anticipate the arrival of deliveries and make sure enough space is available to receive them. These are two critical elements of receiving. Schedule deliveries to avoid peak periods of the day. Employees need to receive and store incoming shipments as soon as they arrive. Always make certain that food products are delivered in clean, well-maintained vehicles that protect the food from contamination and maintain proper product temperature.

Whatever the size of the food establishment, receiving requires:

▲ Prompt handling

▲ Exacting quality control procedures

▲ Trained staff who have good judgment and experience in interpreting:
   ▼ Product specifications
   ▼ Coding
   ▼ Temperature measurement.

Packaged Foods

Heat processing is a common method for preserving foods. Potentially hazardous foods are frequently processed to destroy diseases causing organisms and then placed in a container that is sealed with a hermetic seal. The term **hermetic** refers to a container sealed completely to prevent the entry and loss of gases and vapors. Such a container, since it remains intact, also stops the entry of bacteria, yeasts, molds, and other types of contamination. The most common hermetic containers are rigid metal cans and glass bottles.

Always check cans for:

▲ Leaks

▲ Bulges

▲ Dents

▲ Broken seals

▲ Damage along seams

▲ Rust

▲ Missing labels.

Do not accept cans if they leak or bulge at either end. Swollen ends on a can indicate that gas is being produced inside. This gas may be caused by a chemical reaction between the food and the metal in the container, or it may be caused by the growth of microbes inside the can.

Dents in cans do not harm the contents unless they have actually penetrated the can or sprung the seam. Shipments with many dented cans or torn labels indicate poor handling and storage procedures by the supplier. Figure 4.2 shows some common defects found in cans.

Figure 4.2—Defective Cans

Modified Atmosphere Packaging (MAP) is a process whereby foods are placed in containers and air is removed from the package (Figure 4.3). Different gases, such as nitrogen and carbon dioxide, are then added to the packaged food to preserve it. This technique allows centralized processing and packaging of retail cuts of meat and poultry, thus eliminating entirely the need for processing of the food establishment.

a. pre-cut salad items

b. vacuum packaged fish

Figure 4.3—Examples of MAP Foods

Sous-vide is the French term for "without air." Processors of sous-vide foods seal raw ingredients, often entire recipes, in plastic pouches, then vacuum out the air. They then minimally cook the pouch under precise conditions and immediately refrigerate it. Some processors replace some of the air in the pouches with nitrogen or carbon dioxide. Processing food in this manner eliminates the need for the extreme cold of freezing and the intense heat of canning, thus better preserving taste.

Potentially hazardous foods processed using the sous-vide technique must be kept out of the food temperature danger zone. The lack of oxygen inside the package provides a suitable environment for the growth of *C. botulinum*. **The sous-vide process may not destroy**

these harmful bacteria and does not destroy spores. If the food is not refrigerated and kept out of the temperature danger zone, the spores may germinate and form vegetative bacterial cells. The botulism bacteria can produce toxins that may be fatal if eaten.

The FDA recommends that sous-vide foods be:

▲ Used by the expiration date printed on the package

▲ Refrigerated constantly [safe cold storage temperatures may need to be below 41°F (5°C)]

▲ Be heated according to the time and temperature provided on the package directions.

Food irradiation is a preservation technique used by some food processing industries. This process involves exposing food to ionizing radiation in order to destroy disease-causing microbes and delay spoilage. The FDA has approved food irradiation for a variety of foods including fruits, vegetables, grains, spices, poultry, pork, lamb, and, more recently, ground beef.

The acceptance of irradiated foods has been limited due to consumer fears. However, contrary to many myths, irradiated food is not radioactive and does not pose a risk to consumers. Foods that are processed with irradiation are just as nutritious and flavorful as other foods that have been cooked, canned, or frozen.

Federal law requires that all irradiated food must be labeled with the international symbol for irradiation called a "radura" (Figure 4.4). This symbol must be accompanied by the words "Treated with Irradiation" or "Treated with Radiation."

Irradiation of food can effectively reduce or eliminate pathogens and spoilage microbes while maintaining the quality of most foods. This is a technology that has been proven safe and should be welcomed by consumers as an effective food preservation technique.

Figure 4.4—Radura

Red Meat Products

Primarily red meat and meat products come from cattle (beef), calves (veal), hogs (ham, pork, and bacon), sheep (mutton), and young sheep (lamb). These products are inspected for wholesomeness by officials of the United States Department of Agriculture (USDA) or state agencies. Animals must be inspected prior to and after slaughter to make certain they are free of disease and unacceptable defects. *The USDA also offers voluntary meat grading services. Grades for meat represent the quality or palatability of the meat and are not measures of product safety.* Figure 4.5 contains examples of inspections and grading stamps applied to products approved for use.

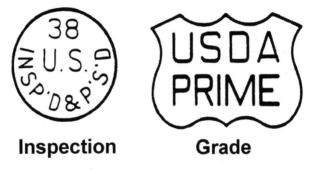

Inspection   Grade

Figure 4.5—USDA Inspection and Grade Stamps for Beef, Veal, and Lamb

Meat and meat products are obtained in several forms such as fresh, frozen, cured, smoked, dried, and canned. **Since meats are potentially hazardous foods, never accept them if there is any sign of contamination, temperature abuse, or spoilage.**

Never accept fresh meat if the product temperature exceeds 41°F (5°C) at delivery. Fresh meat should be firm and elastic to the touch and have characteristic aromas. Off odors are frequently an indicator of spoilage. Sliminess is another characteristic of spoilage and is caused by bacterial growth on the surface of meat. Control of factors that cause spoilage and sliminess also extends the shelf life of meat products.

Frozen meats should be solidly frozen when they arrive at the food establishment. Look for signs of freezing and thawing and refreezing such as frozen blood juices in the bottom of the container or the presence of large ice crystals on the surface of the product. Frozen meats should be packaged to prevent freezer burn.

Move these products quickly from the delivery truck to the freezer for storage.

Game Animals

Game animals are not allowed for sale or service in retail food establishments unless they meet federal code regulations. This ban does not apply to commercially raised game animals approved by regulatory agencies, field dressed game allowed by state codes, or exotic species of animals that must meet the same standards as those of other game animals.

Game animals that are commercially raised for food must be raised, slaughtered, and processed according to standards used for meat and poultry. Common examples of animals raised away from the wild and used for food are farm-raised buffalo, ostrich, and alligator. The USDA inspects the slaughter and processing of this meat in the usual manner.

Some states permit game animals, such as deer, bear, and elk, that have been killed and dressed in the field, to be used in food establishments. Game meat must be dressed soon after the kill to prevent rapid growth of bacteria already present in the meat. Next, the meat must be chilled rapidly, transported in a sanitary manner, and processed in an approved facility. Veterinarians are frequently appointed to inspect meat for contamination that would harm humans.

Some people consider exotic species of animals a delicacy. A distributor providing this type of product (elephant, tiger, monkey, etc.) must meet the same standards as those for domestic wild game. Endangered species may be prohibited by law from being used for food.

Poultry

Poultry is a very popular food item in America. Chicken, turkey, ducks, geese, and other types of poultry are used all year long for roasting, frying, broiling, grilling, and stewing.

All poultry products must be inspected for wholesomeness by USDA or state inspectors. Inspection for wholesomeness involves an examination of the poultry to make certain it is wholesome and not adulterated. The poultry is examined live before slaughter, during evisceration (removal of the internal organs), and during or after packaging. Inspected poultry products carry a USDA seal for wholesomeness on the individual package or on bulk cartons. Examples of inspection and grade stamps for poultry are shown in Figure 4.6.

Usually poultry is graded also for quality. Top quality birds receive a Grade A rating. To earn a Grade A rating, the poultry must have good overall shape and appearance, be meaty, practically free from defects, and have a well-developed layer of fat in the skin.

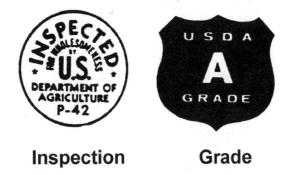

Inspection   Grade

Figure 4.6—USDA Inspection and Grade Stamps for Poultry

Poultry products support the growth of disease-causing and spoilage microorganisms. The intestinal tract and skin of poultry may contain a variety of foodborne disease bacteria, including *Salmonella spp.* and *Campylobacter spp.* The near neutral pH, high moisture, and high protein content of poultry make it an ideal material for bacteria to grow in and on (Figure 4.7).

Poultry products are also vulnerable to spoilage caused by enzymes and spoilage bacteria.

Most spoilage bacteria come from the live bird's skin or intestinal tract. The bacteria grow on the skin of slaughtered birds and inside the birds' carcasses. Spoilage is indicated by meat tissue that is soft, slimy, and has an objectionable odor. Stickiness under the wings is another sign of poultry spoilage. Darkened wing tips on poultry are caused by drying or exposure to freezing temperatures. Poultry that is discolored or has darkened wing tips or sticky skin should be rejected. At the time of delivery, poultry should be packaged on a bed of ice that drains away from the meat as it melts and held at or below 41°F (5°C).

Figure 4.7— Spoiled Poultry

Eggs

Most food establishments, regardless of their size or menu, use eggs in one form or another. Eggs are usually purchased by federal grades, the most common being AA, A, and B (Figure 4.8). Grades for eggs are based on exterior and interior conditions of the egg. Eggs cannot receive a grade when they are dirty, cracked, or broken.

Figure 4.8—USDA Inspection and Grade Stamps for Eggs

The USDA reports that approximately 50 billion eggs are sold in the U.S. each year. *Salmonella enteritidis* bacteria are present in about one percent (approximately 500 million) of the eggs sold. This bacteria enters the yolk of the egg as it is formed inside the hen. The egg shell surface may contain *Salmonella spp.* bacteria, especially if the shell is soiled with chicken droppings. Even if the shell is not cracked, bacteria can enter through the pores in the egg's shell. Raw shell eggs should be clean, fresh, free of cracks or checks, and refrigerated at 45°F (7°C) or below when delivered. (Note: Raw shell eggs can be received at a higher temperature than other potentially hazardous foods). Raw shell eggs should then be stored at 41°F (5°C) or below. The egg, when opened, should have no noticeable odor, the yolk firm, and the white should cling to the yolk. Reject eggs that are dirty or cracked, and remember that washing eggs only increases the possibility of contamination.

An egg product is defined as an egg without its shell. As a safeguard against *Salmonella spp.*, the FDA requires that all egg products, such as liquid, frozen, and dry eggs, be pasteurized to render them *Salmonella*-free. Pasteurized egg products should be in a sealed container and kept at 41° F (5°C). Egg containers should carry labels that verify the contents have been pasteurized.

The shelf life of eggs is limited. Only buy the quantity of eggs the establishment will use in a one- or two-week period.

Fluid Milk and Milk Products

This food group includes milk, cheese, butter, ice cream, and other types of milk products. When receiving milk and milk products, make certain they have been pasteurized. **Pasteurization** destroys all disease-causing microorganisms in the milk and reduces the total number of bacteria, thus increasing shelf-life. All market milk must be Grade A quality. Pasteurization also destroys natural milk enzymes that might shorten the shelf-life of the products.

Milk that is marked "UHT" is pasteurized using ultra-high temperatures and is packaged aseptically. UHT products can be stored safely for several weeks if kept under refrigeration. No refrigeration is required for short storage periods (Longree and Armbruster). Individual creamers are sometimes processed in this manner.

Fluid Milk

Fluid milk is a potentially hazardous food that must be received cold and refrigerated immediately upon delivery. Keep containers of fluid milk and dairy creamers at 41°F (5°C) or below. Individual containers of milk should be clearly marked with an expiration date and the name of the dairy plant that produced it. Check the expiration date of all dairy products before using them.

Cheese

Cheese should be received at 5°C (41°F). It should be inspected to make certain it possesses the proper color, flavor, and moisture content. Cheese should be rejected if it contains mold that is *not* a normal part of the cheese or if the rind or package is damaged.

Butter

Butter is made from pasteurized cream. Since disease-causing and spoilage bacteria and mold may grow in butter, handle the product as a perishable item. The most common type of deterioration in butter is the development of a strong odor and flavor. Check to see that butter has a firm texture, even color, and is free of mold. The package should be intact and provide protection for the contents.

Staphylococcus aureus toxin has been the cause of foodborne disease outbreaks where butter and cheese were involved. *Staphylococcus aureus* does not require as much moisture as other pathogens and can thrive in salty foods which would inhibit other pathogens. Therefore, cheese and butter must be handled carefully and kept out of the temperature danger zone except as required during production.

Fish

Fish includes finfish that are harvested from saltwater and freshwater, and seafood that comes mainly from saltwater. Catfish and trout are two of the most popular types of freshwater finfish. Tuna, snapper, and flounder are saltwater finfish that are very popular with American consumers. Seafood consists of molluscan shellfish and crustaceans. Molluscan shellfish include oysters, clams, mussels, and scallops. Crustaceans include shrimp, lobster, and crab. Oysters, shrimp, catfish, salmon, and a few other types of finfish and seafood are being raised on fish farms using a technique called aquaculture.

Fish and seafood are generally more perishable than red meats, even when stored in a refrigerator or freezer. Fish and seafood are typically packed on self-draining ice to prevent drying and maximize the shelf-life of the product. The slime covering fish and shellfish contains a variety of bacteria that makes them highly susceptible to contamination and microbial spoilage. Fish are also rich in unsaturated fatty acids which are susceptible to oxidation and the development of off-flavors and rancidity.

The quality of fish and seafood is measured by smell and appearance. Fresh finfish should have a mild, pleasant odor and bright, shiny skin with the scales tightly attached. Fish with the head intact should have clear, bulging eyes and bright red, moist gills. The flesh of fresh fish should be firm and elastic to the touch.

Fish must be commercially and legally caught or harvested, except when caught recreationally and approved for sale by the regulatory authority. Ready-to-eat raw, marinated, or partially cooked fish other than molluscan shellfish must be frozen throughout to a temperature of -4°F (-20°C) or below for 7 days in a freezer, or -31°F (-35°C) or below for 15 hours in a blast freezer, before service or sale. Records must be retained that show how the product was handled.

Shellfish must be purchased from sources approved by the Food and Drug Administration and the health departments of states located along the coastline where the shellfish is harvested. Shellfish transported from one state to another must come from sources listed in the *Interstate Certified Shellfish Shippers List*. The reason for requiring tight control over molluscan shellfish is to reduce the risk of infectious hepatitis and other foodborne illnesses that may result from eating raw or insufficiently cooked forms of this product.

When received by a food establishment, molluscan shellfish should be reasonably free of mud, dead shellfish, and shellfish with broken shells. Damaged shellfish must be discarded.

Molluscan shellfish must be purchased in containers that bear legible source-identification tags or labels (Figure 4.9) that have been fastened to the container by the harvester and each dealer that shucks, ships, or reships the shellstock. Molluscan shellfish tags must contain the following information:

▲ The harvester's identification number

▲ The date of harvesting

▲ The harvesting site

▲ The shellfish type and quantity

▲ A statement about leaving the tag on the container until the shellfish is used

▲ Instructions to keep the tag on file for ninety (90) days.

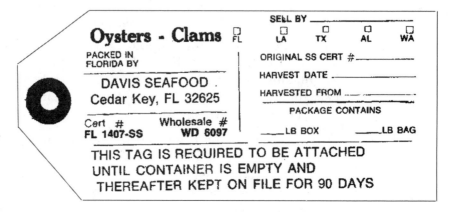

Figure 4.9—Example of Shellfish Tag

A shellstock tag must remain attached to the original container until the container is empty. These tags must be held for at least 90 days after the container is emptied. If seafood is suspected of being the source of foodborne illness, the investigating team can use the tags to determine where and when the product was harvested and processed.

Do not remove shellstock from the shipping container until right before sale or preparation for service, unless it is held on drained ice or in a display case that protects it from contamination. When it is necessary to remove the shellfish from the original, tagged container, only one container of shellfish should be used at a time. Molluscan shellfish caught recreationally *may not* be used or sold in food establishments.

Vegetables and Fruits

Most fruits and vegetables spoil very rapidly. They continue to ripen even after they are picked. Therefore, they may become too ripe if not properly handled. Microorganisms found in water and soil can also cause fruits and vegetables to spoil. Fruits and vegetables hold their top quality for only a few days.

Some products, like wild mushrooms, may only be used if they have been inspected and approved by a mushroom-identification expert who is approved by the regulatory authority. Beware of fresh mushrooms that are packaged in styrofoam trays and covered with plastic shrink wrap. Mushrooms have a high rate of respiration which will quickly use up the oxygen inside the package. Unless holes are poked in the plastic wrap that covers the package to permit oxygen inside, oxygen-free conditions may occur that are favorable for the growth of *C. botulinum* bacteria.

Fresh fruits and vegetables are usually not considered potentially hazardous foods. However, the number of cases of foodborne illnesses linked to these kinds of products have increased in recent years. This is largely due to increased consumption of fresh fruits and vegetables and the emergence of microbes that can cause disease with a low number of organisms. *E. coli* O157:H7, *Shigella spp.*, Hepatitis A, and *Cyclospora spp.* can be infective with only a few cells. Therefore, they do not require a potentially hazardous food to multiply. **Purchase raw fruits and vegetables from approved sources and wash them thoroughly to remove soil and other contaminants before they are cut, combined with other ingredients, cooked, served, or offered for human consumption in a ready-to-eat form.** Though not required, whole raw fruits and vegetables may be washed using cleaners and anti-microbial agents. When these types of chemicals are used, they must meet the requirements in the Code of Federal Regulations (21CFR 173.315). The fruits and vegetables should also be rinsed to remove as much of the residues of these chemicals as possible.

Whole raw fruits and vegetables that will be washed by the consumer before consumption do not need to be washed before they are sold.

Frozen Foods

Frozen products must be solidly frozen when delivered. The temperature of frozen foods can be checked by inserting the sensing portion of a thermometer between two packages. Receiving personnel should also look for signs that the product has been thawed and refrozen. Common signs of thawing and refreezing are large ice crystals on the surface of a frozen food and frozen liquids or juices inside the package. Reject frozen foods that are not solidly frozen or show signs of temperature abuse.

Proper Storage of Food

Merely assigning someone to check incoming shipments is not enough. Besides knowing what to look for upon delivery, employees must also know how and why to safely handle the food products that are left at the establishment. They must move the materials to the desired storage areas quickly, using clean carts and dollies to transport the food.

Stock rotation is a very important part of effective food storage. A **first in, first out (FIFO) method** of stock rotation helps ensure that older foods are used first (Figure 4.10). Product containers should be marked with a date or other readily identifiable code to help food workers know which product has been in storage longest. When expecting food shipments, always make certain the older stock is moved to the front of the storage area to make room for the newly arriving products.

Figure 4.10—Proper Stock Rotation

Types of Storage

The three most common types of food storage areas are the:

▲ Refrigerator

▲ Freezer

▲ Dry storage.

Refrigerated storage is used to hold potentially hazardous and perishable foods for relatively short periods of time, usually a few days. Freezer storage is used to hold foods for longer periods of time, usually a few weeks to several months. Dry storage is typically used to store less perishable items and foods that are not potentially hazardous foods.

Refrigerated storage slows down microbial growth and preserves the quality of foods. Some common types of refrigerated storage equipment are walk-in, reach-in, and pass-through refrigerators. This equipment usually maintains the air temperature in the storage compartment at about 38°F (3° C). Potentially hazardous foods must be stored at 41°F (5°C) or below. Fish and shellfish that are especially vulnerable to spoilage should be stored at colder temperatures ranging from 30°F to 34°F (-1.1° to 1.1°C). Some fruits and vegetables, such as bananas and potatoes, undergo undesirable chemical changes when they are refrigerated. Therefore, although fruits and vegetables are perishable products, not all types of them should be refrigerated. Fresh fruits and vegetables requiring refrigeration should be stored at temperatures between 41°F and 45°F (5° and 7°C). Refrigerators must be equipped with either indicating or recording thermometers to monitor the temperature of the air inside the storage unit.

Freezers are designed to keep foods at 0°F (-18°C) or below. Freezer equipment must also be equipped with indicating or recording thermometers to monitor the temperature of the ambient air inside the unit. If your freezer is not frost free, defrost it regularly to ensure proper operation. Wrap frozen foods and transfer them to the refrigerated storage area until the defrosting process is complete.

Some important procedures for cold storage are listed below:

▲ Rotate refrigerated and frozen foods on a *first in, first out* (FIFO) basis. Store foods in covered containers that are properly labeled and dated.

▲ Store foods in refrigerated and freezer storage areas at least six inches off the floor. Space products to allow the cold air to circulate around them.

▲ Store raw products under cooked or ready-to-eat foods to prevent cross contamination.

Use a *dry storage area* to store foods that are usually packaged in cans, bottles, jars, and bags (see Figure 4.11). These products must be labeled according to federal regulation and come packaged from approved commercial facilities.

Figure 4.11—Dry Storage Area

The dry storage area should have a moderate room temperature of 50° to 70°F (10° to 21°C) and a relative humidity of 50 to 60% to maximize the shelf life of the foods stored there. The quality of some foods is reduced by sunlight. Therefore, windows are not recommended in dry storage areas. If windows are present, they should be blocked out or shaded.

Stored foods in the dry storage area should be on slatted shelves, at least six inches off the floor and away from the wall (Figure 4.12). The six-inch clearance permits thorough cleaning under and behind the shelving. It also discourages insects and rodents from harboring there. Products should be spaced on the shelves so that air can circulate around them.

If it is necessary to transfer bulk items, such as flour, sugar, and grain foods, into other containers, these containers must be constructed of food-grade materials and equipped with tight fitting lids. Scoops and other utensils that are used to remove food from bulk food containers must be constructed of non-toxic, non-absorbent, and easily cleanable material. These utensils must be equipped with handles and stored in a manner that will allow food workers to grip the handle of the utensil without touching the food with their hands.

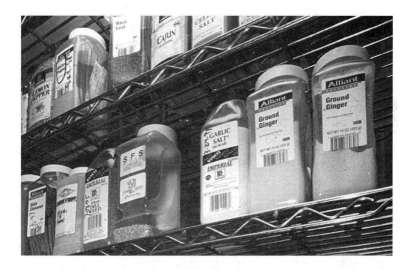

Figure 4.12—Store Dry Foods on Slated Shelving

Make certain the containers are properly coded, or dated, and labeled with the common name of the food. Containers holding food that can be easily and accurately recognized, such as dry pasta, do not need to be labeled.

Do not use toilet rooms, locker areas, mechanical rooms, and similar spaces for storage of food, single-service items, paper goods, or equipment and utensils. Do not expose products to overhead water and sewer lines unless the lines are shielded to interfere with potential drips (Figure 4.13).

Figure 4.13—Contamination of Ready-to-Eat Food

Do not store chemicals, such as cleaning and sanitizing agents, near foods (Figure 4.14). If space permits, store these products in a completely separate area away from the dry storage area. If a separate storage space is not available, store these chemicals in a locked and labeled cabinet. Pesticides must be stored in an area away from food and materials used for cleaning and sanitizing. Always discourage employees from storing their clothing, handbags, and other personal items in dry storage areas.

Figure 4.14—Do Not Store Chemicals Near Food

Storage Conditions for Foods

Meat and meat products come in a variety of forms including fresh, frozen, cured and smoked, dried, and canned. Fresh meat can be stored for up to three weeks at temperatures between 32° and 41°F (0° and 5°C) and a relative humidity between 85 and 90%. Cold temperatures extend the shelf life of red meats by slowing down the growth of bacteria that cause spoilage. The high relative humidity prevents excessive drying and shrinkage.

Frozen meats can be stored for several months when held at 0°F (-18°C) or below. Frozen meats must be wrapped in moisture-proof paper to prevent them from drying out. Packaging for frozen foods should also be strong, flexible, and protect against light.

Poultry can be safely stored at temperatures between 30° and 36°F (-1° and 2°C) for short periods of time. A relative humidity of 75 to 85% is recommended, as excessive humidity causes sliminess due to excessive bacterial growth. Poultry should be wrapped carefully to prevent dehydration, contamination, and loss of quality. Frozen poultry and poultry products can be stored for four to six months when held at 0°F (-18°C) or below.

Whole shell eggs will keep fresh for eight to nine months when stored at 29° to 31°F (-2° to -1°C) and 85 to 90% relative humidity. Egg quality deteriorates rapidly at room temperature. Egg shells are porous and odors can be absorbed very easily. Keep eggs covered and store them away from onions and other foods that have a strong odor. Discard eggs that are dirty or cracked. Always make sure to wash your hands after handling whole shell eggs.

Egg products such as whole eggs, egg whites, and yolks are pasteurized to destroy *Salmonella* bacteria. Fresh egg products should be stored at 41°F (5°C) or below. Store frozen eggs at 0°F (-18°C) or below and keep them frozen until time for defrosting. Refrigerate dried egg products or keep them in a cool, dry place. Once dried eggs have been reconstituted, they are considered potentially hazardous and must be stored out of the temperature danger zone.

Milk is one of the most perishable foods handled in a food establishment. However, properly pasteurized milk that has not been recontaminated and is held at 41°F (5°C) will keep up to ten days or longer. The optimal storage temperature for fluid milk is 33° to 41°F (1° to 5°C), and the shelf life of milk is shortened significantly at higher storage temperatures. For example, milk held at 60°F (16°C) will stay fresh less than one day. Milk also picks up odors from other foods. Therefore, store milk in an area away from onions and other foods that give off odors.

Fish and shellfish are more perishable than red meats, even when refrigerated or frozen. There are many bacteria in the surface slime and digestive tracts of living fish. When a fish is killed, bacteria attack the tissue of the fish. This bacteria lives on cold-blooded fish at low water temperatures. Therefore, they adapt well to the cold and continue to grow even under refrigerated conditions. The shelf life of fresh fish depends on many factors including: species, season of the year, physical condition (whether filleted or dressed), and manner of handling.

Most shellfish are even more perishable than finfish. The shells of live clams and oysters should be closed or close when tapped. Lobsters and crabs should be kept alive until they are cooked or frozen. Otherwise, they lose quality in a day or less.

Food	Freezer	Refrigerator
Meat and Fish		
Bacon (opened)	1-2 months	5-7 days
(unopened)	1-2 months	2 weeks
Deli meat salads	N/A	3-5 days
Fish (fresh)	3-6 months	1-2 days
(cooked)	1 month	3-4 days
(smoked)	4-5 weeks	10 days
Ground beef	3-4 months	1-2 days
Ham (sealed in can)	N/A	6-9 months
Hot dogs (unopened)	1-2 months	2 weeks
(opened)	N/A	5 days
Luncheon meats (unopened)	1-2 months	2 weeks
(opened)	N/A	1 week
Meat pie or casserole	3 months	2-3 days
Meats (fresh)	3-6 months	3-7 days
Meats (ground)	3-4 months	1-2 days
Pork chops	3-4 months	2-3 days
Sausage (fresh)	3-4 months	1-2 days
Poultry and Eggs		
Poultry (cooked)	2 months	1-2 days
Poultry (fresh)	6 months	2 days
Eggs (in shell)	N/A	2 weeks
(hard-cooked)	N/A	1 week
Egg substitutes (unopened)	1-2 months	10 days
(opened)	N/A	3 days
Dairy		
Butter	10 months	2 weeks
Cheese, hard (unopened)	1-2 months	3-6 months
(opened)	N/A	3-4 weeks
(sliced)	N/A	2 weeks
Cottage cheese	N/A	10-30 days
Cream cheese	1-2 months	2 weeks
Margarine (stick form)	12 months	"use by" date
Milk	1 month	"use by" date
Sour cream	N/A	2-4 weeks
Yogurt	N/A	1-2 weeks
Fruit and Vegetables		
Asparagus (fresh)	N/A	2-3 days
(frozen)	1-2 months	1 day
Broccoli (fresh)	N/A	3-5 days
(frozen)	1-2 months	1 day
Cabbage (fresh)	N/A	1 week
Carrots (fresh)	N/A	2 weeks
Cauliflower (fresh)	N/A	1 week
Celery	N/A	7-10 days
Corn (fresh)	N/A	1 day
Fruit (fresh	9-12 months	3-5 days
(dried)	1 year	4 days (cooked)
Lettuce	N/A	1 week
Vegetables (canned, opened)	N/A	1-4 days

Figure 4.15—Recommended Cold Storage Guidelines

(Source: *Washington State University Cooperative Extension Service*)

Fish and shellfish should be used within 24 hours unless they are stored in crushed ice that will drain away from the product as it melts. Products that are stored in ice should be placed *under* cooked or ready-to-eat foods to prevent cross contamination.

The ideal storage conditions for fresh fruits are temperatures between 41° and 45°F (5° and 7°C), a relative humidity of about 80%, and shaded from light. During respiration and ripening, fresh fruits give off carbon dioxide and water. Because of this, they may become wilted and lose flavor. Proper air circulation in the refrigerated storage area is necessary to maintain freshness and firmness. Inadequate ventilation leads to spoilage and deterioration. Regularly inspect stored fruits. Discard any that begin to spoil.

Vegetable storage requires low temperatures and high humidity to preserve texture, tenderness, flavor, color, and nutritive content. To retain top quality, vegetables should be stored between 41° and 45°F (5° and 7°C) with a relative humidity of 85 to 95%. If fresh fruits and vegetables come packed with an airtight film, poke holes in it to allow the contents to breathe. Otherwise, off-flavors and odors may develop.

Preparation and Service

The preparation and service of foods can involve one or more steps. In small food establishments, such as convenience stores, food products are commonly purchased in ready-to-eat forms which are stored until sold to the consumer. Large operations, such as restaurants, supermarkets, and institutional feeding facilities, prepare and serve food in vast quantities. Food production in these larger establishments may span several hours or days. These operations are more complex and may involve many steps.

Regardless of how many steps may be involved in food production and service, foodborne illness prevention requires effective food safety measures that ensure good personal hygiene and avoid cross-contamination and temperature abuse. Food safety strategies aimed at addressing these practices will be discussed in the remainder of this chapter.

Handwashing

Hands, especially the tips of the figures, are known to be significant sources of contamination and cross contamination of foods. This is especially true during the steps of preparation and service.

Prevention of foodborne illnesses begins with good personal hygiene and includes proper handwashing (Figure 4.16). The importance of scrubbing forearms, hands, and nails using soap, running water, and friction cannot be overstated. With the vigorous removal of visible soil, harmful microbes can be washed down the drain.

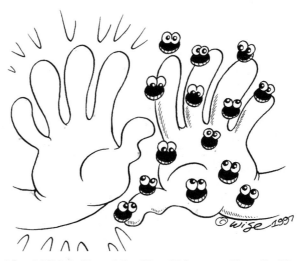

Figure 4.16—Which Hand Do You Want to Handle Your Food?

Avoiding Temperature Abuse

In Chapter 5, the Hazard Analysis Critical Control Point System (HACCP) is presented, and you will identify areas of risk. If you work in a food establishment, you must *know how and when to use a thermometer*. **Temperature and time are the most common critical control points as identified in the HACCP flow charts in preparation and service**. *Monitoring and controlling food temperatures* are extremely effective ways to *minimize the risks of foodborne illnesses*. Thermometers are used for stored, cooked, hot-held, cold-held, and reheated foods. Various types of thermometers are used throughout the food establishment. Some of these instruments were described and shown in Figure 3.3. Before using a thermometer, make sure that it is clean, sanitary, and properly calibrated. The "sensor" portion or probe stem of the thermometer must be inserted into the thickest part of the food.

Freezing

Most harmful bacteria and other microorganisms do not grow at temperatures below 41°F (5°C). However, many spoilage microorganisms grow at much colder temperatures. Therefore, frozen foods must be kept *solidly* frozen.

Sometimes freezing foods can make them safer. Although bacteria are generally not destroyed by freezing, parasites can be killed if frozen at the proper temperature for the proper length of time. Figure 4.17 presents the parasitic destruction guidelines that have been established for raw-marinated and marinated, partially cooked fish.

Food should be frozen throughout to -4° F (-20°C) and held for 7 days in a freezer,

or

Food should be frozen throughout to -31°F (-35°C) using a blast chiller, and held at that temperature for 15 hours.

Figure 4.17—Guidelines for Parasitic Destruction
(Source: *Food Code*)

It is difficult to measure the *internal* temperature of packaged frozen foods. However, it is acceptable to place a thermometer between two fully-frozen packaged foods to gain a reading. Be sure to monitor the temperature of your freezer unit regularly to make sure it is working properly. However, you must understand that measuring the temperature of the freezer is never the same as measuring the actual food temperature.

Thawing

Thawing frozen foods is a common activity in food establishments. The most common and acceptable methods for thawing foods include: in a refrigerator, in a microwave oven followed by immediate cooking, under cold running water, and as part of the cooking process (Figure 4.18).

The preferred method of thawing is in the refrigerator, since foods thawed in this manner would not have the opportunity to

be in the temperature danger zone. This method requires good planning and adequate refrigeration space. It can take two or three days and sometimes longer to thaw large food masses such as turkeys, roasts, and hams. When using this method to thaw foods, you must plan ahead.

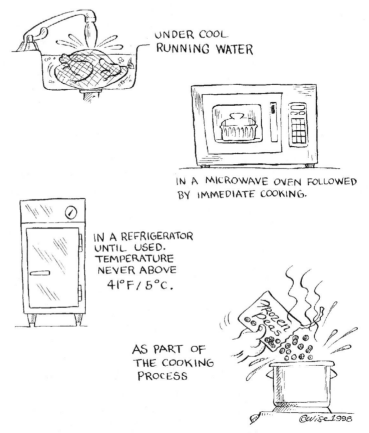

UNDER COOL RUNNING WATER

IN A MICROWAVE OVEN FOLLOWED BY IMMEDIATE COOKING.

IN A REFRIGERATOR UNTIL USED. TEMPERATURE NEVER ABOVE 41°F / 5°C.

AS PART OF THE COOKING PROCESS

Figure 4.18—Safe Ways to Thaw Food

Foods thawed in the microwave must be cooked immediately after thawing. Cooking can be done in the microwave oven or with conventional cooking equipment. Do not thaw foods in a microwave oven and *then* store them in a refrigerator before they are cooked. The thawing process may initiate microbial activity that could reach dangerous levels during refrigerated storage.

Potentially hazardous foods, like turkey and roasts, can be thawed under cool (less than 70°F, 21° C) running water. The thawing time, when using the running water method, should be less than four hours including the time it takes for preparation for cooking or to lower the food temperature to 41°F (5°C) under refrigeration. Thawed portions of potentially hazardous foods should not be allowed to rise above 41°F (5°C). The *Food Code* permits this temperature to increase to 45°F (7°C) if refrigeration equipment that is currently in use is **not** capable of maintaining food at 41°F (5°C). However, the *Food Code* also recommends that equipment that is in use and is not capable of maintaining the food at 41°F (5°C) must be upgraded or replaced to maintain food at proper temperature within 5 years of the regulatory authority's adoption of the *Food Code*. Refer to the guidelines in Figure 4.19 for more information about proper thawing methods

Under no circumstances should foods be thawed at room temperature. Room temperature thawing puts foods in the temperature danger zone—the very thing you don't want to have happen. When foods are thawed at room temperature, the outer surface of the food thaws first and will soon reach room temperature. Microbial growth occurs very quickly at room temperature.

Cold Storage

Most harmful microorganisms start to grow at temperatures above 41°F (5°C). Therefore, it only makes sense to store potentially hazardous and perishable foods in cold storage at 41°F (5°C) or below.

Some cold-loving bacteria can grow at temperatures below 41°F (5°C), but their growth is very slow. You should monitor the temperature of cold held foods routinely throughout the day to ensure that safe temperature requirements are being met.

Refrigerators and cold service bars differ in their capacity to keep foods cool. In all instances, the air inside the refrigerated compartment will need to be a few degrees colder than 41°F (5°C) to ensure that cold foods are held at 41°F (5°C) or below.

Refrigerated potentially hazardous foods must be stored at 41°F (5°C) or less whenever possible. The *Food Code* permits potentially hazardous foods to be stored at 45°F (7°C) or between 45°F (7°C) and 41°F (5°C) if the existing refrigeration equipment in use is **not** capable of maintaining the food at 41°F (5°C). The existing refrigeration equipment that is not capable of maintaining the food at 41°F (5°C) must be upgraded or replaced to maintain food at 41°F (5°C) or less within five years of the regulatory authority adoption of the *Food Code*.

▲ Under refrigeration that maintains the food temperature at 41°F (5°C) or below

▲ Completely submerged under running water

  ▼ at a water temperature of 70°F (21°C) or below

  ▼ with enough water velocity to remove contaminants from the surface of the food

  ▼ for a period of time that does not allow thawed portions of ready-to-eat foods to rise above 41°F (5°C)

  ▼ for a period of time that does not allow thawed portions of a raw animal food requiring cooking to be in the temperature danger zone for more than a total time of 4 hours

▲ As a part of the cooking process

▲ Use any procedure (e.g., microwave oven) that thaws a portion of frozen ready-to-eat food that is prepared for immediate service in response to an individual consumer's order.

Figure 4.19—Guidelines for Thawing Food
(Source: *Food Code*)

Refrigerated, ready-to-eat, potentially hazardous foods include deli meats, potato and macaroni salads, chicken, and seafood salads, cooked shrimp, and similar items. The *Food Code* recommends that when these products are prepared and held refrigerated for more than 24 hours, they be clearly marked at the time of preparation to indicate the date by which the food must be consumed. The acceptable storage time for these types of products is seven calendar days when held at 41°F (5°C) or below and four calendar days when the food is held at 45°F (7°C) or below. The storage time begins with the day the food is prepared.

Frozen, ready-to-eat, potentially hazardous foods should be consumed within 24 hours after thawing whenever possible. When large amounts of food are removed from the freezer, it should be marked to indicate the date by which the food must be consumed. Food must be consumed:

▲ seven calendar days or less after the food is removed from the freezer, minus the time before freezing that food is held refrigerated, if the food is maintained at 41°F (5°C) or less before freezing

▲ four calendar days or less after the food is removed from the freezer, minus the time before freezing that the food is held refrigerated, if the food is maintained at 45°F (7°C) or below.

When storing raw potentially hazardous foods like meats [Figure 4.20(a)], be sure that they are maintained at less the 41°F (5°C). Usually the temperature of the refrigerator should be at least 3-5°F (1.5-3°C) below the desired cold holding temperature. Fish and seafood products should be maintained on ice [Figure 4.20(b)] to prevent the growth of disease-causing bacteria and spoilage bacteria.

a. fresh meat

b. fresh fish

Figure 4.20—Cold storage of raw potentially hazardous foods

When storing read-to-eat potentially hazardous foods like prepared salads and luncheon meats [Figure 4.21(a)] be sure that the refrigerator unit can maintain a safe cold holding temperature. This may be more difficult in open refrigerators that do not contain doors. Many open refrigerator units have a "safe load line" [Figure 4.21(b)]. This line indicates the level at which foods should be stored below to ensure that the food is at the proper temperature.

a. prepared salads and luncheon meats

b. The safe load line

Figure 4.21—Cold Storage of RTE Foods

Cooking

You must expect raw foods, especially those of animal origin, to contain harmful microorganisms. Foods such as meat, poultry, fish, seafood, eggs, and unpasteurized milk should not be prepared and served raw or rare. Establishments that choose to serve raw foods increase their risk of having a foodborne illness. Practices such as using raw eggs to make a Caesar salad, or serving raw oysters on the half shell, increase the risk of a foodborne illness. Raw animal foods need to be cooked to the proper temperatures to be safe.

The purpose of **cooking** is to make food more palatable by changing its appearance, texture, and aroma. Cooking also heats the food and destroys harmful microorganisms that may be found in and on the product. The destruction of disease-causing microorganisms is a phenomenon involving a direct relationship between time and temperature. For example, requirements for cooking a particular food item should include a final internal temperature as well as a prescribed length of time at that temperature.

Most foods are cooked using stoves, conventional ovens, and microwave ovens. Because heat transfer can be different depending on the heating source, final temperature requirements have been set for conventional oven cooking and microwave cooking. Cooking guidelines for various kinds of potentially hazardous foods are presented in Figure 4.22.

Rare beef roasts require the least internal temperature requirements. This is because the contamination is on the surface of the large roast. When an internal temperature of 130°F (54°C) for 121 minutes or 145°F (63°C) for 3 minutes is reached, the surface temperature of the food is much higher. This heat will destroy any pathogens that may have been on the surface of the roast. Animal foods such as eggs, fish, and beef (other than roasts) should be cooked to an internal temperature of at least 145°F (63°C) and held at that temperature for at least 15 seconds before serving.

Cook ground beef and pork and game animal products to 155°F (68°C) and hold for at least 15 seconds before serving. Pork and game animals need a higher cooking temperature due to the possible presence of the trichinosis worm and other parasites.

Food Type	Minimum Internal Temperature	Minimum Time Held at Internal Temperature Before Serving
Beef Roast (rare)	130°F (54°C)	121 min.
Eggs, Beef and Pork (other than roasts), Fish	145°F (63°C)	15 sec.
Ground Beef and Pork, Game Animals	155°F (68°C)	15 sec.
Beef Roast (medium), Pork Roast, and Ham	145°F (63°C)	3 min.
Poultry, Stuffed meats	165°F (74°C)	15 sec.

Note: When microwave cooking, heat raw animal foods to a temperature of 165°F (74°C) in all parts of the food.

Figure 4.22—Cooking Guidelines for Potentially Hazardous Foods
(Source: *Food Code*)

Poultry and stuffed meats should be cooked to an internal temperature of 165°F (74°C) and held for at least 15 seconds before serving.

Casseroles and other foods that contain a combination of raw ingredients such as meat and poultry must be cooked to a final temperature that coincides with the highest risk food. In this case, 165°F (74°C) is required to destroy pathogens that may be found in the poultry. Food mixtures, such as chili and beef stew, must be cooked to 165°F (74°C) to assume proper destruction of disease-causing agents.

While not usually potentially hazardous foods, fruits and vegetables that are cooked should reach an internal temperature of 140°F (60°C) or above.

When cooking foods in the microwave oven, the distribution of heat is often uneven. To distribute the heat more evenly you must stir frequently and rotate the food. The current *Food Code* requires raw animal foods cooked in a microwave oven to be heated to 165°F (74°C) in all parts of the food. Some state codes require microwave heated foods to reach a temperature 25°F (14°C) higher than the temperature required when using conventional equipment. This helps to provide an extra margin of safety. For example, chicken breasts cooked in a conventional oven should reach an internal temperature of 165°F (74°C), whereas poultry cooked in a microwave should reach 165°F + 25°F (74°C + 14°C) or 190°F (88°C). Allow microwaved foods to stand covered for 2 minutes before serving to allow heat to disperse more evenly.

Measure the *internal temperature* of foods while they are *being cooked*. Ideally, the internal temperature should be measured in the geometric center of the food. A commonly used practice in the food industry is inserting the appropriate thermometer or thermocouple probe into the thickest part of the food mass. This will give you an accurate reading of the internal temperature of the product.

Proper cooking is extremely important for the preparation of safe food. If a potentially hazardous food is cooked to a safe temperature, bacteria and other harmful organisms will be destroyed. Bacterial spores and some toxins may not be destroyed by normal cooking temperatures. Therefore, proper handling of these foods is required at every stage of the food flow.

Cooling

Improper cooling is the number one cause of foodborne illness in food establishments. Foods are in the temperature danger zone during cooling and there is no way to avoid it. After proper cooking, potentially hazardous foods need to be cooled from 140°

to 41°F (60° to 5°C) as rapidly as possible. The *Food Code* recommends that hot foods, not used for immediate service or hot display, be cooled from 140° to 70°F (60° to 21°C) within 2 hours, and from 70° to 41°F (21° to 5°C) within an additional 4 hours. Industry standards commonly recommend that hot foods be cooled from 140° to 41°F (60° to 5°C) in four hours. Industry often chooses a more stringent time guide than those suggested in the *Food Code*.

Large quantities of food and foods that have a thick consistency take a long time to cool. For instance, it can take 72 hours or more for the center of a 5-gallon stockpot of steamed rice to cool down to 41°F (5°C) when taken hot from the stove and placed in a refrigerator.

Foods must pass through the temperature danger zone as quickly as possible. Figure 4.23 lists some of the more commonly used methods for reducing cooling time.

▲ Use containers that facilitate heat transfer (stainless steel)

▲ Transfer food into shallow pans that will allow for a product depth of 3″ or less

▲ Transfer food into smaller containers

▲ Stir food while cooling

▲ Place containerized food in an ice water bath

▲ Stir food in a container placed in an ice water bath

▲ Use cooling paddles to stir the food

▲ Add ice directly to a condensed food.

Figure 4.23—Methods to Reduce Cooling Time for Food

Photographs of the three most commonly used methods of cooling are presented in Figure 4.24. As it was with cooking, you must monitor times and temperatures for foods that are cooling. When large masses of food are broken down into smaller portions, cooling is more rapid.

For example, a large turkey can be sliced into smaller pieces, or creamed corn can be transferred into shallow pans to facilitate cooling. Foods cool faster in food-grade metal containers than in food-grade plastic containers.

When using shallow pans to cool liquid food, the pan should be only 4 inches high. The depth of foods cooled in a shallow pan should be less than 3 inches for liquids (soup) and less than 2 inches for viscous products (chili or beef stew) [Figure 4.24(b)].

Always use a thermometer to verify that foods are cooling properly. Never assume that any one method is working without checking the temperature and time that foods take to cool. Always depend on the thermometer reading with any of the methods you use to cool food.

a. Ice bath   b. Place food in shallow   c. Use a cooling paddle
                     pans 2-3" deep   Courtesy of KatchAll Industries International

Figure 4.24—Approved Cooling Methods

Cooling time is greatly reduced when foods are stirred. Foods stirred every 10-15 minutes cool far more rapidly than unstirred foods. An ice water bath also helps to remove heat from foods.

An ice water bath involves putting the container of hot food inside a larger container and surrounding it with ice to promote cooling. It is sometimes desirable to combine cooling methods, such as stirring a food after putting the container into an ice water bath.

You may also fill a large pot with ice and water. Then, insert the container of hot food into the ice water and stir the product. Keep the water level below the rim of the food container. In some instances, ice from a potable water supply can be added as a "final ingredient" in preparing hot items from condensed foods. Large quantities of soup or broth can be prepared in this way.

Hot-holding, Cold-holding, Reheating

All potentially hazardous foods that have been cooked, cooled, and then reheated must be held hot at 140°F (60°C) or above. **Cold-**

holding is holding potentially hazardous foods which are to be consumed cold at 41°F (5°C) or below. **Hot-holding** is holding potentially hazardous foods above 140°F (60°C) during transportation and delivery to any site away from the primary preparation and service areas.

Figure 4.25—Hot Holding—Store Hot Held Foods with the Utensils in the Food

Proper holding temperatures slow down or prevent the growth of harmful microorganisms. The key is to keep foods out of the temperature danger zone (140° to 41°F, 60° to 5°C). All potentially hazardous foods meant to be eaten hot, which have been cooked and then cooled, must be reheated to at least 165°F (74°C) within two hours. Foods can be heated more quickly when heated in smaller quantities or by using pre-heated ingredients. Stir food frequently during reheating to reduce the time needed to reheat. Foods should only be reheated once, and uneaten portions of a reheated food should be discarded.

Serving Safe Food

When you serve food, always practice good personal hygiene. Start by wearing a clean uniform and hair restraint. Food, or surfaces that may come in contact with food, should not be touched with hands. Hold utensils only by the handles and do not touch beverage glasses by the outside or inside rim. Handle plates and bowls by the bottom or outer rim (Figure 4.26). Finally, wash your hands after handling dirty tableware and utensils. Never dry your hands on your apron. Always use a single-service towel.

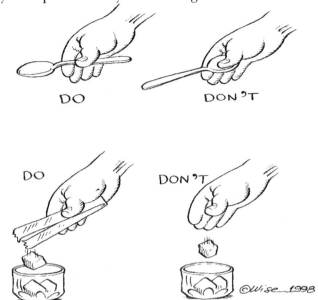

Figure 4.26—Handle Ice and Eating Utensils Properly

Self-Service Bar

Self-service bars are very popular. They offer convenience and a wide range of selections for consumers (Figure 4.27). A properly

Figure 4.27—Self-Service Bar

installed sneeze guard protects the food from contamination by your customers and is required in most jurisdictions (Figure 4.28).

Figure 4.28—Don't Forget the Sneeze Guard

Keep potentially hazardous foods on self-service bars hot at 140°F (60°C) or above and cold at 41°F (5°C) or below. Monitor food holding temperatures frequently in these operations. Never serve raw animal products at a self-service bar.

Always use clean sanitized utensils in a self-service bar. Monitor the bar and replace any utensils that become contaminated. Hold utensils by the handle. Do not allow hand contact with foods in containers.

Keep utensils in the food between use. Resting utensils on saucers beside containers is not an acceptable practice. Only one utensil should be used for each food item.

Self-service bars tend to get messy because of spills. *Clean the station* frequently throughout the day, to reduce the potential for cross contamination. Assign a properly trained food worker to monitor and maintain a self-service bar or buffet bar.

Self-service plates and bowls should only be used once to avoid contaminating food on the buffet or salad bar. Give your customers a clean plate or bowl each time they return to the salad bar. Beverage cups and glasses may be reused to get refills.

Temporary and Mobile Food Facilities

The popularity of temporary and mobile food facilities such as street fairs, festivals, and mobile carts increased rapidly during the past decade (Figure 4.29). The public patronizes these events in increasing numbers. In addition to the opportunity for community involvement, commercial and non-commercial organizations are finding it profitable to sell food at temporary food facilities.

Mobile **Temporary**

Figure 4.29—Examples of Temporary and Mobile Food Facilities

In the beginning, temporary facilities served food items such as hamburgers and hot dogs. These foods required minimal preparation. They were taken from a cooler, placed on a grill, and cooked to order.

Depending on where you are, the food served by today's temporary facilities might consist of anything from BBQ ribs, calamari, gyros, and pizza to sushi and shish kebabs. Today's sophisticated appetites demand foods which frequently involve risky preparation methods, such as heating, cooling, and reheating.

Consumers also insist on freshly prepared food, prompting vendors to transport raw ingredients to their make-shift booths to chop, shred, and assemble without access to the sanitary facilities that would be standard in a permanent food establishment.

The more extensive the menu or more complex the preparation, the higher the risk, and the more demanding public health requirements become to reduce these dangers. Vendors should contact the local regulatory agency in their area to obtain a copy of the rules that govern temporary and mobile operations.

Protecting the food and food preparation equipment from contamination is the function of the structure. A temporary food stand should have:

▲ An overhead covering

▲ An enclosed area except for the serving windows and an entry door

▲ A source of hot and cold potable running water for handwashing, cleaning, and sanitizing.

Your menu and method of food preparation will determine your selection of equipment and utensils. Food contact surfaces must be designed for effective cleaning and sanitizing. They must also be constructed of materials that are non-toxic. Cooking equipment must be capable of heating potentially hazardous foods to the required internal temperature. For example, hamburgers and other ground beef should be cooked to an internal temperature of 155°F (68°C); poultry to 165°F (74°C); and fish and other meats to 145°F (63°C). Foods to be reheated must reach an internal temperature of 165°F (74°C) within two hours. **Do not use slow cooking devices such as crockpots, steam tables, and sterno to heat foods. These devices may never reach temperatures high enough to kill harmful bacteria.**

Foods that require refrigeration must be cooled to 41°F (5°C) as quickly as possible and held at that temperature until served. Cold foods should be at 41°F (5°C) or below when delivered to the temporary facility. If for some reason it is necessary to cool foods down, use an ice water bath or place the food in shallow pans and refrigerate. Do not forget that proper food temperatures are your most important control measure against foodborne illnesses. Minimizing the time between preparation and service is another good practice to use in order to keep food safe.

An adequate handwashing facility must be provided, and strict handwashing procedures must be observed to keep food safe. Disposable gloves can provide an additional barrier to contamination. However, gloves are no substitute for handwashing.

Good food handling practices are an important line of defense in temporary food facilities. Raw foods must be kept and handled separate from ready-to-eat foods. It is best to use tongs, napkins, disposable gloves, or other tools to handle food.

Temporary food facilities should use disposable utensils for food service. If it is necessary to use multi-use utensils, they must be washed and sanitized using a four-step process:

1. Washing in hot, soapy water

2. Rinsing in clean water

3. Chemical sanitizing

4. Air drying.

A detailed explanation of this process is presented in Chapter 7, Cleaning and Sanitizing Operations. It is a good idea to have additional cleaned and sanitized utensils on hand to accommodate peak periods when there is not adequate time to perform the dishwashing process. Rinse and store the wiping cloths used in the operation in a pail of sanitizer solution. The concentration of the sanitizer must be checked regularly to make certain it remains at the required strength. The strength of the sanitizer can be measured using either a chemical test kit or test strips. Always replace the solution whenever it falls below the required minimum concentration.

Garbage and paper wastes should be placed in containers that are lined with plastic liners and equipped with tight-fitting lids. Keep foods and wastes covered to avoid attracting insects, rodents, and other pests. Always store pesticides away from food, and be sure to follow the manufacturer's instructions if you apply them. Clean and sanitize all food contact surfaces that may have been contaminated when the pesticides were applied.

Along with the fun and potential profit, food vendors should be aware of the potential public health problems with food served by temporary or mobile food service operations. Improvised facilities, large volumes of food, severe weather conditions, and untrained food workers can make keeping foods safe and sanitary a real challenge. Temporary events frequently serve large numbers of people which makes it necessary to prepare large quantities of food in advance. This increases the risk of contamination and makes proper cooking, cooling, and holding even more critical. Always consult your local food protection program on the specifics of your operation.

Home Meal Replacement

Home meal replacement has become a multi-billion dollar business for retail food establishments. A loosely defined term, home meal replacement is most often used to refer to high-quality meals prepared away from home and eaten at home.

The move toward prepared food has supermarkets selling meal solutions instead of just ingredients. Supermarkets and restaurants are competing head-to-head for the "heat and eat" business that has become very popular with today's consumer.

Home replacement meals come in the "ready-to-cook," "ready-to-heat," and "ready-to-eat" varieties. All three varieties are designed to save time and effort for families that are too frazzled to cook at the end of the day.

The "ready-to-cook" line, including foods such as beef and chicken kebabs with the peppers and onions already cut up, has found its way into most grocery stores' meat or frozen food departments.

"Ready-to-heat" is a much broader segment of the market. It includes the fully cooked hams, turkey roast, or breaded chicken cutlets found in supermarket meat sections. Many supermarkets have expanded their deli sections to include items that can be conveniently reheated, such as pizzas, rotisserie chickens; and hot-tray choices such as macaroni and cheese, meat loaf, and mashed potatoes. Several stores also stock packaged pre-cooked eggs rolls, burritos, chicken dinners, and other foods in the cold food case.

"Ready-to-eat" foods come from fast-food takeout counters and pizzas shops—many of which offer family meals deals that serve four to six people. Consumers stop by on the way home from the office and buy complete meals ready to eat right out of the box.

With home replacement meals, retailers must deliver quality, consistency, and safe food—and have it ready when the customers are there. By following specific food safety guidelines, retailers can ensure that the meals their customers buy to consume at home are free from the germs that cause foodborne illness. Hot foods must be cooked to the required temperature for the required amount of time and held out of the temperature danger zone. Cold foods must be chilled quickly to control microbial growth and held at 41°F (5°C) or below during storage and display.

Home replacement meals should be labeled so customers understand how to keep the product safe when they take it home. Warn them against keeping the food in the car while they do more shopping or keeping it at room temperature when they get home. Pamphlets and brochures stuffed in to bags or stapled to the front of bags can help educate consumers about safe handling of home replacement meals. In addition, it is a good idea to establish "best-before" dates and codes for these foods. Retailers must be able to show they've done everything in their power, including written documentation and following industry standards, to make the food safe.

Summary

The investigation of the *E. coli* O157:H7 outbreak presented in the case study at the beginning of this chapter concluded the *E. coli* bacteria was most likely introduced into the coleslaw by contaminated cabbage. This provides an example of how food can be contaminated when it arrives at a food establishment. Cabbage is grown on the ground where it can be contaminated by animal feces that contain *E. coli* O157:H7 bacteria. In this particular case, the cabbage was shipped in net bags. This could have allowed the cabbage to become contaminated during transport. Food workers did not wash the cabbage prior to shredding even though it was quite soiled. This likely allowed the *E. coil* O157:H7 to be introduced into the coleslaw. At least one other restaurant in the area purchased cabbage from the same supplier. However, workers at that establishment did wash the cabbage after removing the leaves. No problems were reported at that establishment.

Prevention measures that would help prevent similar types of foodborne outbreaks are washing vegetables and fruits that are to be served raw, proper personal hygiene, and proper cleaning and sanitizing of equipment and utensils to prevent cross contamination.

Not all food managers or supervisors actually will select or purchase products. However, knowledge of the rules, regulations, and procedures of receiving and storage is expected of all who are responsible for food safety.

To ensure that all foods are safe for human consumption, purchase your food supplies from approved sources and suppliers who comply with all laws concerning food processing and labeling. Schedule deliveries for the off peak times when someone is available to inspect incoming food supplies. Make sure that food is free from filth or spoilage, in sound condition, and is at the proper temperature. If the product has been damaged or does not meet the standards, reject the delivery. The supplier should be responsible for replacing or giving credit for rejected material.

Move delivered materials into designated storage areas quickly. Storage serves as a temporary link between receiving and preparation. Storage facilities must be capable of preserving the sanitary quality of the food stored there.

Store foods in their original containers or packages whenever possible. When food is removed from the package or container in which it was delivered, it must then be stored in a clean, covered container that is properly labeled and dated. Proper stock rotation using the FIFO system is necessary to ensure that stocked items will be used within a reasonable period of time.

Provide adequate "dry storage" for dry items such as canned goods. For highly perishable items, refrigerator or freezer storage is necessary. Chilled storage is required for perishable and potentially hazardous foods such as dairy products, meat, poultry, eggs, fish, seafood, fruits, and vegetables. Store these products out of the temperature danger zone. Freezer storage, capable of holding products in a frozen condition, must be provided for all frozen foods.

To ensure the maintenance of all potentially hazardous food at the required cold temperatures during storage, refrigerators and freezers must be equipped with a numerically scaled thermometer. The thermometer should measure temperature in increments no greater than 2°F (1°C). The sensor of the temperature-measuring device must be located to measure the air temperature in the warmest part of a mechanically refrigerated unit. The temperature-sensing device should also be located to allow easy viewing of the device's temperature display. The temperature of food should be measured to verify that holding temperatures are correct.

To prevent the possibility of cross contamination, always store raw or unwashed foods below ready-to-eat foods and foods that will not receive any additional cooking. Do not store food containers on the floor or under any exposed or unprotected sewer or water lines.

Keep the storage areas, including the floors, walls, and ceilings, clean. Make sure that the shelving units in these areas are clean and in good repair. Spilled foods should be cleaned up immediately and removed from the area.

Errors that occur during the preparation and service of food are the leading causes of foodborne illness in food establishments. To ensure safe food, proper food handling is essential.

Safe food preparation and service include good personnel hygiene and taking measures to avoid cross contamination. It is critical to know when and how frequently hands and other food contact surfaces should be cleaned and sanitized. Monitoring and controlling food temperatures during preparation are important for destroying microorganisms and preventing them from growing in the food. Keeping foods out of the temperature danger zone as much as possible and cooking them properly are both very important considerations. Foods should only be allowed in the temperature danger zone for a short time during thawing, heating, and cooling activities. The three main problems in foodservice, time and temperature abuse, cross contamination, and poor personal health and hygiene practices of food workers, must be prevented during the flow of food. Preparation and service are critical processes in your food establishment because they are the last steps you take before the consumer eats your food.

Case Study 4.1

David Jones is manager of the Great American Cafe. As part of a newly implemented self-inspection program, David performs an inspection of the walk-in refrigerator. During the inspection, David notices that turkeys for tomorrow's dinner are not covered during thawing, and they are stored directly above several washed heads of lettuce. Other food items are stored in covered containers that are not labeled or dated. David also notices that boxes of produce have been stacked closely together on the floor of the walk-in cooler, and the thermometer is hanging from the condensing unit.

1. What food safety hazards exist in the walk-in refrigeration unit at the Great American Cafe?

2. Which of these hazards might result in food contamination and spoilage?

3. What should be done to correct the problems that Mr. Jones observed during his inspection?

Case Study 4.2

Metro Market is a combination convenience store and delicatessen. Deliveries are received between 9:00 a.m. and 3:00 p.m., Monday through Friday. Store employees are frequently too busy to check the deliveries in and transfer products to approved storage facilities during the noon rush. Therefore, food and nonfood items are held in a secured area off the receiving dock until someone is available to process them.

1. What food safety hazards exist at Metro Market?

2. If you were manager of Metro Market, what would you do to improve receiving and storage activities?

Case Study 4.3

In the mid 1990s, health departments in Washington, California, Idaho, and Nevada identified nearly 600 people with culture-confirmed *E. coli* O157:H7 infection. Many of the victims reported eating at Chain A restaurants during the days preceding onset of symptoms. Of the patients who recalled what they ate in a Chain A restaurant, a large percentage reported eating regular-sized hamburger patties. Chain A issued a multi-state recall of unused hamburger patties.

The *E. coli* O157:H7 bacteria were isolated from 11 batches of patties from Chain A. These patties had been distributed to restaurants in all states where the illness occurred. Approximately 20% of the patties connected with the outbreak were recovered during a recall.

A team of investigators from the Centers for Disease Control and Prevention identified five slaughter plants in the United States and one in Canada as the likely sources of carcasses used in the contaminated meat. The animals slaughtered in domestic slaughter plants were traced to farms and auctions in six western states. No one slaughter plant or farm was identified as the source of contamination.

E. coli O157:H7 can be found in the intestines of healthy cattle, and can contaminate meat during slaughter. Slaughtering practices can result in contamination of raw meat with these bacteria. In addition, the process of grinding beef can transfer the disease-causing agents from the surface of the meat to the interior. Therefore ground beef can be internally contaminated.

1. What would be the likely impact of this outbreak on consumer confidence?

2. What was the impact of this outbreak on food safety?

The answers to the case studies are provided in Appendix A.

Discussion Questions (Short Answer)

1. Purchase specifications are important to a food establishment's operation. Why?

2. Why should buyers purchase foods only from approved sources that comply with all applicable food laws?

3. Discuss the differences between grading and inspection services.

4. What range of temperatures should food temperature measuring devices be able to measure?

5. Why should cans with swollen ends be rejected and sent back to the supplier?

6. What are typical signs of spoilage for red meats, poultry, and fish?

7. What is the meaning of FIFO?

8. Why should raw products be stored below cooked or ready-to-eat foods during storage?

9. Describe safe procedures for storing, cleaning, and sanitizing agents and pesticides.

10. Why should products be located at least 6 inches off the floor during storage?

11. Why is it important for a food worker to wash his/her hands?

12. Why is it important to **not** serve raw animal foods?

13. Why do safe temperature requirements differ when using a conventional oven as opposed to the microwave oven?

14. Suggest some ways to decrease the cooling time for a 5-gallon container of rice.

Quiz 4.1 (Multiple Choice)

1. Which of the following is **not** a rule that should be closely followed when purchasing food?

   a. Foods prepared in a private home may not be used or offered for human consumption in a retail food establishment.

   b. Buyers should only purchase food that is safe, wholesome, and from an approved source.

   c. Avoid the use of commercially raised game animals as meat and poultry items.

   d. Only buy meat and poultry that has been inspected by USDA or state agency officials.

2. The best method for measuring the temperature of frozen food products is by:

   a. Inserting the sensing probe into the center of a frozen food package until the recorded temperature stabilizes.

   b. Inserting the sensing probe between two packages of frozen foods until the recorded temperature stabilizes.

   c. Measuring the ambient temperature of the frozen food compartment of the delivery vehicle.

   d. Looking for signs of freezing and thawing, such as large ice crystals and frozen juices in the bottom of the box.

3. Frozen foods should **not** be accepted at a food establishment if:

   a. They have large ice crystals on the surface.

   b. There are frozen juices on the bottom of the package.

   c. If the temperature is above 32°F (0°C).

   d. All of the above.

4. Which of the following foods should **not** be rejected upon delivery?

    a. Fresh fish that has dull, sunken eyes, and soft flesh.

    b. Poultry with darkened wing tips and soft flesh.

    c. Canned fruit with small amounts of surface rust on the lid of the can.

    d. Fresh beef products that are delivered at 45°F (7°C).

5. Which of the following statements about fish and seafood is **false**?

    a. Fish and shellfish are less likely to spoil than red meat and poultry.

    b. Quality in fish and shellfish is measured by smell and appearance.

    c. Fish that may be eaten raw must be commercially frozen prior to consumption.

    d. Molluscan shellfish tags must be kept for 90 days from the date the container is emptied.

6. Which of the following storage practices should prompt a manager to take corrective action?

    a. Products in the dry storage area are being rotated on a first-in, first-out stock basis.

    b. Foods stored in the walk-in freezer are stored on slatted shelves that are 6 inches above the floor.

    c. Raw poultry is stored above potato salad in the walk-in refrigerator.

    d. Cleaning and sanitizing agents and pesticides are stored in a locked and labeled cabinet in the dry food storage area.

7. Pork roasts should be cooked to an internal temperature of at least _____ for 15 seconds to be considered safe.

    a. 140°F (60°C)

    b. 145°F (63°C)

    c. 155°F (68°C)

    d. 165°F (74°C)

8. Which of the following is the preferred method for thawing potentially hazardous foods?

    a. In the microwave oven.

    b. At room temperature.

    c. In the refrigerator.

    d. Under 70°F (21°C) running water.

9. Hot foods should be held at _____ or above and cold foods should be held at _____ or below.

    a. 165° ; 41°F (74° ; 5°C)

    b. 165° ; 32°F (74° ;0°C)

    c. 140° ; 41°F (60° ; 5°C)

    d. 140° ; 32°F (60° ; 0°C)

10. Poultry and stuffed meats should be cooked to an internal temperature of _____ for 15 seconds to be considered safe.

    a. 140°F (60°C)

    b. 145°F (63°C)

    c. 155°F (68°C)

    d. 165°F (74°C)

11. Ground beef meats should be cooked to an internal temperature of _____ for 15 seconds to be considered safe.

    a. 140°F (60°C)

    b. 145°F (63°C)

    c. 155°F (68°C)

    d. 165°F (74°C)

12. Regardless of the type of food, all potentially hazardous foods that have been cooked and cooled need to be reheated to an internal temperature of _____ within two hours to be considered safe.

    a. 140°F (5°C)

    b. 145°F (63°C)

    c. 155°F (68°C)

    d. 165°F (74°C)

13. According to the FDA *Food Code*, foods that are cooked and then cooled must be cooled in which of the following ways:

    a. From 140° to 41°F (60° to 5°C) in 12 hours.

    b. From 140° to 41°F (50° to 5°C) in 8 hours.

    c. From 140° to 70°F (60° to 21°C) in 2 hours, and from 70° to 41°F (21° to 5°C) in an additional 4 hours.

    d. From 140° to 70°F (60° to 21°C) in 4 hours, and from 7°F to 41°F (21° to 5°C) in an additional 2 hours.

14. All foods that are to be held cold must be held at _____ or below:

    a. 41°F (5°C)

    b. 50°F (10°C)

    c. 70°F (21°C)

    d. 0°F (-18°C)

15. All foods that are to be held hot must be held at _____ or above:

    a. 70°F (10°C)

    b. 98°F (37°C)

    c. 120°F (49°C)

    d. 140°F (60°C)

Answers to the multiple choice questions are provided in Appendix A.

References/Suggested Readings

Cichy, R. F. (1994). *Quality Sanitation Management*. Educational Institute of the American Hotel and Motel Association. East Lansing, MI.

Educational Foundation of the National Restaurant Association (1998). *Servsafe*, National Restaurant Association. Chicago, IL.

Food and Drug Administration (1999). *1999 Food Code*. United States Public Health Service. Washington, DC.

Kotchevar, L. H., and R. Donnelly (1994). *Quantity Food Purchasing*. Macmillan Publishers. New York, NY.

Longrèe, Karla and G. Armbruster (1996). *Quantity Food Sanitation.* John Wiley and Sons, Inc. New York, NY.

Rue, N., and National Assessment Institute (1994). *Handbook For Safe Food Service Management.* Prentice Hall. Englewood Cliffs, NJ.

Thayer, David W., et.al. (1996). *Radiation Pasteurization of Food.* Council for Agricultural Science and Technology Issue Paper, No. 7. April, 1996.

Selected Web Sites

Gateway to Government Food Safety Information
www.foodsafety.gov

United States Department of Agriculture (USDA)
www.usda.gov

Centers for Disease Control and Prevention (CDC)
www.cdc.gov

United States Food and Drug Administration (FDA)
www.fda.gov

The American Egg Board
www.aeb.org

America Meat Institute
www.meatami.org

The National Cattlemen's Beef Association
www.ncanet.org

The National Chicken Council
www.eatchicken.com

U.S. Poultry and Egg Association
www.poultryegg.org

Information about the Production Marketing Association
www.fightbac.org/about/pma.html

USDA Foods Safety and Inspection Service (FSIS)
www.fsis.usda.gov

USDA/FDA Food and Nutrition Information Center
www.nal.usda.gov/fnic/

United States Environmental Protection Agency (EPA)
www.epa.gov

Partnerships for Food Safety Education
www.fightbac.org

The Food Marketing Institute
www.fmi.org

National Restaurant Association
www.restaurant.org

5

THE HAZARD ANALYSIS CRITICAL CONTROL POINT (HACCP) SYSTEM: A SAFETY ASSURANCE PROCESS

Many Employees are Not Thankful for Thanksgiving Luncheon

At least 40 employees of a small company became ill after a catered Thanksgiving luncheon. The most common symptoms experienced by the victims were diarrhea, cramps, bodyaches and chills, nausea, and fever.

A local food establishment had catered the meal. Investigators from the local regulatory agency asked the food workers to describe the food production practices that led up to suspect meal. The workers indicated that four 18 to 20-pound frozen turkeys were placed in the walk-in cooler to thaw on November 17. The turkeys were cooked on November 20. No product temperatures were taken, but the turkeys had pop-up thermometers that indicated when the birds were done. Immediately following cooking, the turkeys were covered with foil and placed in the walk-in cooler. On the day of the company's luncheon, the turkeys were reheated, sliced, and placed into hot holding units for transport to the luncheon. The food workers indicated that product temperatures were not routinely monitored during cooling and re-heating.

Learning Objectives

After reading this chapter, you should be able to:

▲ Recognize the usefulness of the HACCP system as a food protection tool

▲ Recognize the types of potentially hazardous foods that commonly require a HACCP system to ensure product safety

▲ Identify the steps involved in implementing a HACCP system

▲ Define:

   1. Hazard

   2. Hazard Analysis

   3. Critical Control Point

   4. Critical Limit

▲ List hazards (risk factors) related to each product analyzed

▲ Assess hazards (risk factors) in order of severity

▲ Identify points in the flow of food to be monitored

▲ Describe evaluation of the process

▲ State measures used to correct potential problems

▲ Identify data required to provide documentation for review and problem solving

▲ Apply the HACCP system to analyze and protect food items from contamination during processing, preparation, and service.

Essential Terms

Critical Control Point (CCP)	Hazard Analysis Critical Control Point (HACCP) system
Critical limit	Monitoring
HACCP plan	Risk
Hazard	Verification
Hazard analysis	

The Problem

The challenge of reducing the number of cases of foodborne illness has made it necessary to reevaluate our nation's food safety system. The food industry and food regulatory agencies are being confronted with a number of new challenges, including:

▲ New germs that cause foodborne illness

▲ The changing nature of our global food supply

▲ New techniques for processing and serving food

▲ Changing eating habits of consumers

▲ A growing number of people who are at increased risk of experiencing foodborne illness.

The number of foods imported into the U.S. increases every year. In addition, the number of new methods used to process and prepare foods continues to grow. As appetites change, new items are produced and different foods are grown and imported. This challenges the government's ability to effectively monitor food safety. In addition, governmental agencies at all levels are having to cut back their staff and services due to increased costs and taxpayer mandates to decrease spending. This makes it necessary to shift more and more of the responsibility for food safety to food establishments managers and employees. Those individuals who are engaged in food production and service are required to assume greater responsibility for ensuring the safety of the food they produce.

The size and scope of the food industry exert great pressures on the FDA, USDA, and state and local food safety programs. The FDA has identified more than 30 thousand food manufacturers and 20 thousand food warehouses that process and store products. In addition, the retail portion of the food industry consists of more than one million establishments and employs a work force of over 12 million people. For these reasons, the Hazard Analysis Critical Control Point (HACCP) food safety system is being recommended as the best method for ensuring food safety in retail establishments. This system has been used by food processors for many years to monitor and protect food from contamination.

The Solution

In a HACCP food safety system the greatest amount of attention is placed on food and how it is handled during storage, preparation, and service. Less emphasis is placed on general sanitation issues such as the cleanliness of floors, walls, and ceilings. A sanitary environment is important for safe food production, but food can still be contaminated by employees if they do not use proper food handling techniques, practice good personal hygiene, or control food temperature properly.

The HACCP system helps food managers identify and control potential problems *before* they happen. Whether the HACCP system is employed in restaurants, retail food stores, institutions, health care facilities, or other food service operations, the primary goal is always the same—production of safe and wholesome food.

A HACCP food safety system is most effective when tailored to the specific needs of the retail food establishment. HACCP should **not** be viewed as a "one size fits all" program. It is designed to provide flexibility to the food establishment when controlling the hazards that cause foodborne illness. However, the HACCP system must be compatible with the products sold, the clients served, and the facilities and equipment used during food production.

The HACCP System

Food safety programs of the past have reacted to problems and corrected hazardous conditions after they happened. The HACCP system, on the other hand, is designed to anticipate and control problems *before* they happen. HACCP is the preferred approach to retail food safety because it provides the most effective and efficient way to ensure that food products are safe (*Food Code*). The HACCP system offers two additional benefits over conventional inspection methods.

First, **the HACCP system enables food managers to identify the foods and processes that are most likely to cause foodborne illnesses**. When a potential problem is identified, the food establishment can initiate procedures to reduce or eliminate the risk of foodborne illness and then monitor to make sure the procedures are being followed.

Second, the HACCP system more accurately describes the overall condition of the establishment. Traditional assessments provide only an account of the establishment's condition at the time of the evaluation. Those violations present at the time of the audit are noted, but they may not relate directly to food, time, and temperature. The HACCP system enables food managers and regulatory agency personnel to track food handling practices over a period of time.

Unlike audits that sometimes focus on aesthetic or non-critical items, **the HACCP approach is based on controlling time, temperature, and specific factors that are known to contribute to foodborne disease outbreaks.** Records produced in conjunction with the HACCP system provide a comprehensive source of information about the events that occurred during all stages of food production.

The first priority of the HACCP system is to ensure the safety of the potentially hazardous foods on the menu. You will begin by developing HACCP flowcharts for recipes which allow you to follow the flow of food (see Chapter 4) from start to finish. In the course of developing HACCP flowcharts, you will identify "high risk" activities that occur during food production and which might contribute to a foodborne illness.

You then ask, "What can be done to control these high-risk activities to reduce the risk of foodborne illness?"

HACCP recipes are guides for food workers during production.

Food managers and supervisors can use the flow charts and production logs to double check for product safety.

HACCP records also assist health department personnel as they perform routine inspections of your establishment.

If a foodborne disease outbreak occurs, HACCP records are very useful to the investigators as they attempt to determine the cause of the illness.

Food managers and supervisors must learn how to effectively develop, implement, and maintain a HACCP system. This begins with knowing about and understanding the factors that affect food safety. There are several publications available, including an excellent guide in the *Food Code*, that can help you learn more about the HACCP system. Consult the suggested readings at the end of this chapter for a partial list of these publications.

The Seven Steps in a HACCP System

The basic structure of a HACCP system consists of seven steps as listed in Figure 5.1. While each step is unique, they all work together to form the basic structure of an effective food safety program. A more detailed description of each step will be presented in the remainder of this chapter.

1. Hazard Analysis.

2. Identify the Critical Control Points (CCP) in food preparation.

3. Establish Critical Limits (thresholds) which must be met at each identified Critical Control Point.

4. Establish procedures to monitor CCPs.

5. Establish the corrective action to be taken when monitoring indicates that a Critical Limit has been exceeded.

6. Establish procedures to verify that the HACCP system is working.

7. Establish effective record keeping that will document the HACCP system.

Figure 5.1—Seven Steps in a Hazard Analysis Critical Control Point System
(Source: *Food Code*)

Step 1—Hazard Analysis

The first step in a HACCP system is hazard analysis. This involves identifying *hazards* that might be introduced to food by certain food production practices or the intended use of the product. Hazard analysis starts with a thorough review of your menu or product list to identify all of the potentially hazardous foods you serve. As you learned in Chapter 2, *Hazards to Food Safety*, potentially hazardous foods have properties that support rapid bacterial growth and can cause the food to become unsafe. Potentially hazardous foods include:

▲ Meats

▲ Dairy products

▲ Poultry, eggs

▲ Cooked foods such as beans, pasta, rice, potatoes

▲ Cut cantaloupe and raw seed sprouts.

All of these foods are commonly found in food establishments.

Hazards may be biological, chemical, or physical in nature. These hazards are frequently introduced into the food by people, poor food-handling practices, and/or contaminated equipment. Some foods may be naturally contaminated, as in the case of raw meats, poultry, fish, and unwashed fruits and vegetables.

Biological Hazards

Food is not produced in a *sterile (microorganism free)* environment. Therefore, it can easily be contaminated by biological hazards such as bacteria, viruses, fungi, and parasites. During hazard analysis, look for steps in food production where foods may become contaminated, and microorganisms may survive and multiply. Knowledge of biological hazards sources and the foods they most frequently contaminate is essential to an effective HACCP plan. Refer to Chapter 2, *Hazards to Food Safety*, to review biological hazards and the conditions they need for growth and survival.

Do not forget: Moderate hazards can be severe if transmitted to high-risk individuals such as the very young, frail elderly, or those with weakened immune systems.

Chemical Hazards

Chemical hazards are substances that are either naturally present or are added to food during production. When constructing a HACCP flowchart, consider the possibility of chemical contamination. Train food workers in the proper handling and storage of chemicals. This is essential to your overall food safety program. Reduce the chance of chemical contamination by purchasing food items and additives from approved sources. For more information about chemical contamination refer to Chapter 2 of this book.

Physical Hazards

Physical contaminants in food can cause injury to the consumer. Glass, metal shavings, a food worker's personal property (jewelry, false fingernails, and hair pins), toothpicks, and pieces of worn equipment are examples of physical agents that may accidentally enter food during production and service. A HACCP plan helps foodservice managers and supervisors to identify potential sources of physical hazards and provide safeguards to protect foods from this form of contamination.

During the hazard analysis step, you should also estimate risk. **Risk is the probability that a condition or conditions will lead to a hazard.** Some of the factors that influence risk are the:

▲ Type of customers served

▲ Types of foods on the menu

▲ Nature of the organism

▲ Past outbreaks

▲ Size and type of food production operations

▲ Extent of employee training.

Each type of operation and each food establishment pose different levels of risk to the consumer. **Hazards that pose little or no risk, or are unlikely to occur, need not be addressed by your HACCP system.**

Focus on your menu's most hazardous foods first. Once your HACCP system has been developed, implemented, and evaluated for these foods, move on to the next most hazardous foods. Within several months, you should be able to include all of the potentially hazardous foods on your menu under the HACCP system.

Review the menu items in Figure 5.2 and circle the potentially hazardous foods.

Foods that are not potentially hazardous in the sample menu on page 177, do not support the growth of harmful bacteria. Fruits, such as orange juice, grapefruit half, and strawberries, all have a pH less than 4.6. Disease-causing bacteria will not grow in these acidic conditions. Melons, such as watermelon and cantaloupe, that have a pH greater than 4.6, are potentially hazardous.

Products such as shredded wheat, toast, and angel food cake will not support the growth of disease-causing bacteria due to their low water activity.

Tossed salad is generally not considered a potentially hazardous food. However, it is a ready-to-eat food that is not cooked. It must be handled carefully to prevent contamination and control microbial growth. For example, salad ingredients and other ready-

Breakfast

Orange Juice	Apple Juice	Grapefruit Half	Strawberries
Oatmeal	Cream of Wheat	Shredded Wheat	Raisin Bran
Scrambled Eggs	Bacon	Sausage Links	Hash Browns
French Toast	Sausage Gravy	Cheese Omelet	Pancakes
Belgian Waffle	Breakfast Burrito	White/Wheat Toast	Egg Beaters™

Lunch

Apple Sauce	Pasta Salad	Potato Salad	Spinach Salad
Chili	Navy Bean Soup	Clam Chowder	Vegetable Soup
French Fries	Hamburgers	Pork Tenderloin	Fish Fillet
Chicken Fillet	Ham & Cheese	Chicken Wings	Corned Beef & Swiss

Dinner

Tossed Salad	Cobb Salad	Cottage Cheese	Tuna Salad in Tomato
Baked Potato	Broccoli & Cheese	Wild Rice	Melon Balls
Country Fried Steak	Turkey/Dressing	Liver & Onions	Meat Loaf
Frozen Yogurt	Chocolate Brownie	Cherry Pie	Angel Food Cake
Coffee	Iced Tea	Milk	Soft Drinks

Figure 5.2—Sample Menu

to-eat foods may be contaminated by a virus from the food workers' hands. In the case of viral infections, it takes a very small dose of harmful virus to cause an illness. When handling this type of food, handwashing and temperature control are important safeguards. The hazard identification phase of hazard analysis is made easier by asking a series of questions which are appropriate to each step in the flow of the food. The questions presented in Appendix D help you to identify potentially hazardous foods and possible problems. Select preventive measures that will protect foods based on the hazards and risks you have identified.

Hazard identification and risk estimation provide a logical basis for determining which hazards are significant and must be addressed in the HACCP plan. **The severity of a hazard is defined by the degree of seriousness of the consequences, should it become a reality**. Hazards that involve low risk and are not likely to occur need not be addressed in the HACCP plan.

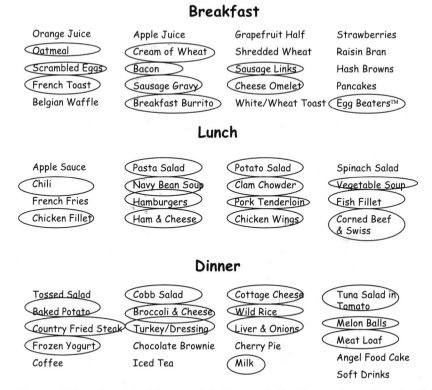

Figure 5.3—Sample Menu with Potentially Hazardous Foods Circled

Personal experience, facts generated by foodborne illness investigations, and information from scientific articles can be useful when estimating the approximate risk of a hazard. When estimating risk, it is important to separate *safety* concerns from *quality* issues. For instance, as a quality issue, perishable foods *spoil* quickly when stored at temperatures in the food temperature danger zone. It is a *safety concern* because foodborne disease investigations show that potentially hazardous foods left in the temperature danger zone too long can cause foodborne illness.

The last phase of the hazard analysis step involves establishing preventive measures. After the hazards have been identified, you must consider what preventive measures, if any, can be employed for each hazard.

Preventive measures include:

▲ Controlling the temperature of the food

▲ Cross contamination control

▲ Good personal hygiene practices

▲ Other procedures that can prevent, minimize, or eliminate an identified health hazard (e.g., limiting the amount of time a potentially hazardous food spends in the temperature danger zone).

More than one preventive measure may be required to control certain types of hazards. For example, the transfer of *Salmonella* bacteria from raw chicken to a ready-to-eat food, such as lettuce or carrots, can be controlled by proper cleaning and sanitizing of equipment and utensils and handwashing practices. In other situations, more than one hazard may be controlled by a specific preventive measure. By cooking a food to 165°F (74°C), many different types of microbes are eliminated, thus making the food safe.

Traditionally, HACCP deals only with preventive measures that can be easily monitored. Since food temperature and time can be easily monitored, they are the preventive measures used most often in HACCP.

Step 2—Identify Critical Control Points (CCPs)

The next step in creating a **HACCP** system is to identify the **critical control points (CCPs)** in food production. **A critical control point is an operation (practice, preparation step, or procedure) in the flow of food which will prevent, eliminate, or reduce hazards to acceptable levels.** A critical control point provides a *kill step* that will destroy bacteria or a *control step* that prevents or slows down the rate of bacterial growth.

Some examples of CCPs are:

▲ Cooking, reheating, and hot-holding

▲ Chilling, chilled storage, and chilled display

▲ Receiving, thawing, mixing ingredients, and other food handling stages

▲ Product formulation (e.g., reducing the pH of a food to below 4.6 or the A_w to .85 or below)

▲ Purchasing seafood, MAP foods, and ready-to-eat foods where further processing would not prevent a hazard, from approved sources.

The most commonly used CCPs are cooking, cooling, reheating, and hot/cold holding. Cooking and reheating to proper temperatures will destroy bacteria, whereas proper cooling, hot-holding, and cold-holding will prevent or slow down the rate of bacterial growth.

The *Food Code* also recognizes specific food handling and sanitation practices (such as proper thawing methods), prevention of cross contamination, and certain aspects of employee and environmental hygiene as CCPs. These latter types of practices are more difficult to measure, monitor, and document. Therefore, many food establishment operators prefer to think of them as "standard operating procedures" (SOPs) or "house policies" rather than CCPs.

For the purpose of this textbook, CCPs are considered to be operations that involve:

▲ **Time**

▲ **Temperature**

▲ **Acidity**

▲ **Purchasing and receiving procedures related to:**

  ▼ **Seafood**

  ▼ **Modified atmosphere packaged foods**

  ▼ **Ready-to-eat foods where a later processing step in the food flow would not prevent a hazard**

▲ **Thawing of ready-to-eat foods where a later processing step in the food flow would not prevent a hazard.**

SOPs include:

▲ **Good employee hygiene practices (i.e., handwashing)**

▲ **Cross contamination control (i.e., keeping raw products separate from cooked and ready-to-eat foods)**

▲ **Environmental hygiene practices (i.e., effective cleaning and sanitizing of equipment and utensils).**

Figure 5.4—CCPs and SOPs

There must be at least one critical control point in the production process to qualify it as a HACCP food safety system, and the *critical control point must be monitored and controlled to guarantee the safety of the food.*

Identification of critical control points *begins with a review of the recipe for the potentially hazardous ingredients* and the *development of a flow chart* for the recipe. The flow chart will track the steps in the food flow from receiving to serving. The specific path that a food follows will be slightly different for each product. However, some of the more common elements in the flow of food include:

▲ Purchase of products and ingredients from sources inspected and approved by regulatory agencies

▲ Receiving products and ingredients

▲ Storage of products and ingredients

▲ Preparation steps which may involve thawing, cooking, and other processing activities

▲ Holding or display of food

▲ Service of food

▲ Cooling food

▲ Storing cooled food

▲ Reheating food for service.

Controlling the temperature of food products throughout the flow of food is the most essential preventative measure for ensuring a safe food. However, **time can also be used as an important measure**. As you learned in Chapter 2, the growth of microbes is dependent on both time and temperature. It commonly takes 4 hours or more in the temperature danger zone for bacteria to multiply to levels where they will cause foodborne illness.

There are situations during food preparation and service where foods will be allowed to enter the temperature danger zone. For example, some foods that are cooked and held hot can not be held at 140°F (60°C) or above without lowering the culinary quality of the food. When fried chicken is held at 140°F (60°C) or above, it quickly dries out and loses it quality. To prevent this, a food establishment can hold the cooked chicken at less than 140°F (60°C). Rather than controlling the temperature of the food, the emphasis is placed on controlling the amount of time the food is in the danger zone. When time is used as a CCP, don't allow more than four hours from production to consumption.

To help you develop a HACCP plan that is effective for your facility, refer to the Critical Control Point Decision Tree in Appendix D. By answering each of the questions in the decision tree you can identify critical control points that will influence the safety of the product. Apply the decision tree at each step in the food production process where a hazard has been identified.

Step 3—Establish the Critical Limits (Thresholds) Which Must Be Met at Each Critical Control Point

At this step in the process, consider what should be done to reduce the hazards risk to safe levels. Set critical limits to make sure that each critical control point effectively blocks a biological, chemical, or physical hazard. **Critical limits** should be thought of as the upper and lower boundaries of food safety. When these boundaries are exceeded, a hazard may exist or could develop. The critical limit should be as specific as possible, such as "ground beef or pork must be heated to an internal temperature of 155°F (68°C) or greater for at least 15 seconds." A well-defined critical limit makes it easier to determine when the limit has **not** been met.

Each CCP has one or more critical limits (Figure 5.5) to monitor to assure that hazards are:

▲ Prevented

▲ Eliminated

▲ Reduced to acceptable levels.

Each limit relates to a process that will keep food in a range of safety by controlling (Figure 5.5):

▲ Temperature

▲ Time

▲ The ability of the food to support the growth of infectious and toxin-producing microorganisms.

"This is Critical!"

Critical Limit	Boundaries of Food Safety
Time	Limit the amount of time food is in the temperature danger zone during preparation and service processes to 4 hours or less.
Temperature	Keep potentially hazardous foods at or below 41°F (5°C) or at or above 140°F (60°C). Maintain specific cooking, cooling, reheating and hot-holding temperatures.
Water Activity	Foods with a water activity (A_w) of .85 or less do **not** support growth of disease causing bacteria.
pH (acidity level)	Disease-causing bacteria do **not** grow in foods that have a pH of 4.6 or below.

Figure 5.5—Criteria Most Frequently Used for Critical Limits

"Now what do I see?"

To be effective, each critical limit should be based on information from food regulatory codes, scientific literature, experimental studies, and food safety experts. **You must be able to easily measure or observe a critical limit.**

Recipes and sample flow charts for three foods served in food establishments are presented in Figures 5.6, 5.7, 5.8, and 5.9. Review these carefully paying particular attention to the steps from source through the preparation process to service.

Ingredients	Amount	Servings 25	50	100
Fresh broccoli	Heads	2.0	4.0	8.0
Red onions	Lbs.	0.5	1.0	2.0
Fresh mushrooms	Lbs.	1.0	2.0	4.0
Sprial pasta	Lbs.	1.0	2.0	4.0
Black olives, sliced	Cups	0.5	1.0	2.0
Italian salad dressing	Cups	0.5	1.0	2.0
Ranch-style dressing	Cups	0.5	1.0	2.0

Figure 5.6—Sample Recipe and HACCP Flow Chart for Pasta Salad

Pre-preparation

SOP-1 1. Wash your hands prior to beginning food preparation.

SOP-2 2. Clean and wash broccoli, onions, and mushrooms under cool running water. Use immediately in the pasta salad or cover and refrigerate at 41°F (5°C) until needed.

3. Pre-chill the salad dressings by storing them at 41°F (5°C) until ready for use in the pasta salad.

Preparation

4. Cook pasta in boiling water until tender. Drain and quick-chill in ice water. Drain after chilling.

5. Combine pasta, broccoli, onions, mushrooms, and olives.

6. Combine the salad dressings in another bowl and mix well.

7. Mix the salad with the dressing.

CCP-1 8. Quick-chill salad. Cool the salad to 41°F (5°C) within 4 hours. Monitor and record the temperature of the product at least once every hour. Cover and hold for service at 41°F (5°C) or below.

Service

CCP-2 9. Maintain temperature of finished product at 41°F (5°C) or below during entire serving period. Monitor and record internal temperature of the product every 30 minutes. Maximum holding time is 4 hours.

10. Discard any portion of the product that is placed into service but is not served within the 4-hour holding time.

Storage

CCP-3 11. Store the chilled product in a covered container that is properly dated and labeled. Refrigerate at 41°F (5°C) or below.

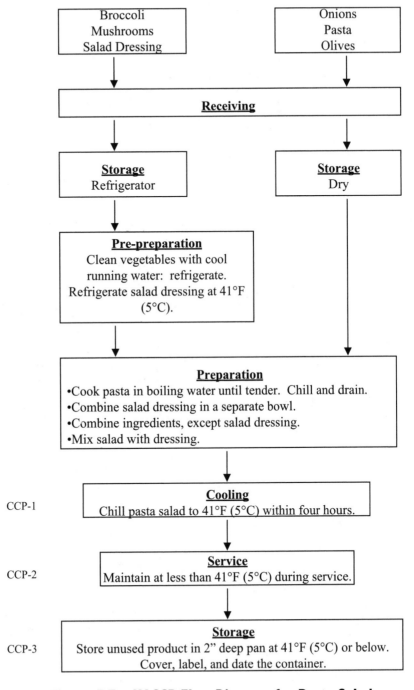

| Broccoli
Mushrooms
Salad Dressing | Onions
Pasta
Olives |

Receiving

Storage
Refrigerator

Storage
Dry

Pre-preparation
Clean vegetables with cool running water: refrigerate. Refrigerate salad dressing at 41°F (5°C).

Preparation
- Cook pasta in boiling water until tender. Chill and drain.
- Combine salad dressing in a separate bowl.
- Combine ingredients, except salad dressing.
- Mix salad with dressing.

CCP-1   **Cooling**
Chill pasta salad to 41°F (5°C) within four hours.

CCP-2   **Service**
Maintain at less than 41°F (5°C) during service.

CCP-3   **Storage**
Store unused product in 2" deep pan at 41°F (5°C) or below. Cover, label, and date the container.

Figure 5.7—HACCP Flow Diagram for Pasta Salad

Ingredients	Amount	Servings		
		25	**50**	**100**
Ground beef	Lbs.	6	12	24
Onion, diced	Cups	2	4	8
Green pepper, diced	Cups	1	2	4
Celery, diced	Cups	1	2	4
Bread crumbs	Cups	3	6	12
Pasteurized eggs, frozen	Cups	1	2	4
Catsup	Cups	1	2	4
Homogenized milk	Cups	½	1	2
Pepper	Cups	½	1	2
Salt	Cups	½	1	2

Figure 5.8—Sample Recipe and HACCP Flow Chart for Meat Loaf

Pre-preparation

SOP-1   1. Wash hands before beginning food production.

SOP-2   2. Thaw ground beef and pasteurized eggs under refrigeration at 41°F (5°C).

SOP-3   3. Clean and rinse vegetables with cool running water and cut as directed. Use immediately in the recipe or cover and refrigerate at 41°F (5°C) or below.

Preparation

4. Combine all ingredients and blend in a properly cleaned and sanitized mixer.

5. Place the mixture in a shallow loaf pan that does not exceed 2-1/2" in depth.

CCP-1   6. Bake the meat loaf uncovered to an internal temperature of 155°F (68°C) for a minimum of 15 seconds in a preheated conventional oven at 350°F (177°C) or a convection oven at 325°F (163°C) for approximately one hour.

7. Remove from oven and allow each cooked loaf to cool at room temperature for 15 minutes. Place loaves in a shallow service pan and slice into 3-ounce portions.

Holding/Service

CCP-2   8. Maintain temperature of product at 140°F (60°C) or higher during service period. Record temperature of unused product every 30 minutes.

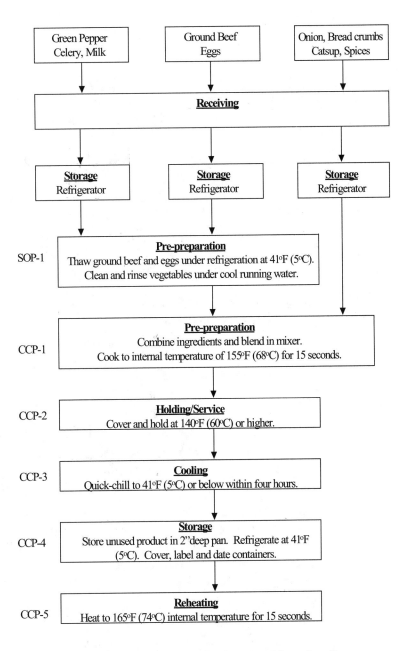

| Green Pepper Celery, Milk | Ground Beef Eggs | Onion, Bread crumbs Catsup, Spices |

Receiving

| **Storage** Refrigerator | **Storage** Refrigerator | **Storage** Refrigerator |

SOP-1
Pre-preparation
Thaw ground beef and eggs under refrigeration at 41°F (5°C).
Clean and rinse vegetables under cool running water.

CCP-1
Pre-preparation
Combine ingredients and blend in mixer.
Cook to internal temperature of 155°F (68°C) for 15 seconds.

CCP-2
Holding/Service
Cover and hold at 140°F (60°C) or higher.

CCP-3
Cooling
Quick-chill to 41°F (5°C) or below within four hours.

CCP-4
Storage
Store unused product in 2"deep pan. Refrigerate at 41°F
(5°C). Cover, label and date containers.

CCP-5
Reheating
Heat to 165°F (74°C) internal temperature for 15 seconds.

Figure 5.9—Flow Chart for Meat Loaf

(Adapted from: *HACCP for Food Services—Recipe Manual and Guide, 1993*)

Storage

CCP-3   9. Cool in shallow pans with a product depth not to exceed 2 inches. Quick-chill the product from 140°F to 70°F (60°C to 21°C) within 2 hours and from 70°F to 41°F (21°C to 5°C) within 4 hours.

CCP-4   10. Store the meat loaf in a covered container at a product temperature of 41°F (5°C) or below.

Reheating

CCP-5   11. Remove from refrigeration and heat in a preheated conventional oven at 350°F (177°C). Heat until all parts of the product reach an internal temperature of 165°F (74°C) for 15 seconds.

         12. Reheat, and then discard unused product.

Step 4—Establish Procedures to Monitor CCPs

In each food establishment, someone should be responsible for **monitoring critical control points**. To monitor, make observations and measurements to determine whether a critical control point is under control. This will expose when a critical control point has exceeded its critical limit.

"You have to monitor it!"

For example, monitoring tells you whether or not the internal temperature of poultry has reached 165°F (74°C) or above for 15 seconds. **The risk of foodborne illness increases when a critical control point is not met. Monitoring is a critical part of a HACCP system and provides written documentation that can be used to verify that the HACCP system is working properly.**

Time, temperature, pH, and *water activity* are the critical limits most commonly monitored to ensure that a critical limit is under control. If a product or process does not meet critical limits, immediate corrective action is required before a problem occurs.

Monitoring can be performed either continuously or at predetermined intervals in the food production process. Continuous monitoring, such as that used to manage the temperature and timing of a cook-chill operation, is performed using temperature recording charts. Continuous monitoring is always preferred because it provides on-going feedback that can be used to determine when critical limits have been exceeded.

If a critical limit cannot be monitored continuously, set up specific monitoring intervals that can accurately indicate hazard control. If early monitoring indicates the process is very consistent, do not monitor as often.

Monitoring must be "doable." More frequent monitoring catches problems earlier and provides more options for correction (i.e., reheat instead of discard food).

Observations of cross-contamination control, employee hygiene compliance, and product formulation control may also be incorporated into the monitoring system. A record of actions, times, temperatures, and any departure from critical limits provides a history for that item. Once the flow is established, measurements can be recorded on a flowchart. Have employees place their initials by the information they record.

Food workers responsible for monitoring CCPs must know how to accurately monitor critical control points and record the information accurately in data records. They must also be given the tools (temperature measuring devices, pH meters, etc.) needed to take measurements and record their findings. Another important part of monitoring is making employees aware of the harm that can occur if data is faked or not collected as required.

Monitoring is one of the most important activities in a HACCP system. An operation that identifies critical control points and establishes critical limits without having a monitoring system in place has not actually implemented a HACCP system. Critical limits without monitoring them are meaningless.

Step 5—Establish the Corrective Action to Be Taken When Monitoring Shows That a Critical Limit Has Been Exceeded

TAP TAP TAP

"What do you do when it is outside the limit?"

Serious problems can occur when critical limits are not met. If you detect that a critical limit was exceeded during the production of a HACCP monitored food, correct the problem immediately. The flow of food should not continue until all CCPs have been met.

First, determine what went wrong. Next, choose and apply the appropriate corrective action. For example, if the temperature of the BBQ pork on your steam table is not at 140°F (60°C) or higher, check the steam table to make sure it is working properly and will keep food hot. At the same time, put the pork on the stove and reheat it rapidly to 165°F (74°C). The pork should be discarded if you suspect it has been in the temperature danger zone for four hours or more.

Additional corrective actions include having employees measure the temperature of the product at more frequent intervals and stir the pork to ensure the even distribution of heat throughout the product. Record the additional steps and verify that the critical limit is met using the revised system. Taking immediate corrective action is vital to the effectiveness of your food safety system.

Step 6—Establish Procedures to Verify That the HACCP System Is Working

"Have I told you lately that it's working?"

The sixth step in the HACCP system is verifying that your system is working properly. The **verification** process typically consists of two phases. First, you must verify that the critical limits you have established for your CCPs will prevent, eliminate, or reduce hazards to acceptable levels. Second, you must verify that your overall HACCP plan is functioning effectively. Once critical limits at each CCP are met, minimal sampling of the final product is needed. However, frequent reviews of HACCP-based food flow plans and records are necessary to ensure product safety. Figure 5.10 contains guidelines that can assist you in determining how and when to implement HACCP plan verification procedures as well as the kind of information your verification reports should containThe HACCP system should be reviewed and, if necessary, modified to accommodate changes in:

▲   your clientele (i.e., more high-risk clients)

▲   the items on your menu (addition of potentially hazardous foods or substitution of low-risk foods for high-risk foods)

▲   the processes used to prepare HACCP products.

Your management team should review and evaluate the establishment's HACCP program at least once a year, or more often if necessary.

Verification procedures may include:

▲ Initiation of appropriate verification inspection schedules

▲ A review of the HACCP plan

▲ A review of the CCP records

▲ A review of departures from critical limits and how they were corrected

▲ Visual inspection of food production operations to determine if CCPs are under control

▲ A random sample collection and analysis

▲ A review of the critical limits to verify if they are adequate to control hazards

▲ A review of written record of verification inspections which certifies compliance with the HACCP plan or deviations from the plan and the corrective action taken

▲ Validation of the HACCP plan, including on-site review and verification of flow diagrams and Critical Control Points

▲ A review of the modifications made to the HACCP plan.

Verification inspections should be conducted under the following conditions:

▲ Routinely and unannounced to ensure that selected CCPs are under control

▲ If it is determined that intensive coverage of a specific food is needed because of new information concerning food safety

▲ When foods prepared at the establishment have been linked to a foodborne illness

▲ When established criteria have not been met

Figure 5.10—Guidelines for Determining How and When to Implement HACCP Plan Verification Procedures

(Source: *Food Code*)

▲ To verify that changes have been implemented correctly after a HACCP plan has been modified.

Verification reports should contain the following information:

▲ Existence of the HACCP plan and person(s) responsible for administering and updating plan

▲ The status of records associated with CCP monitoring

▲ Direct monitoring data of the CCP while in operation; certification that monitoring equipment is properly calibrated and in working order; deviations; and corrective actions noted

▲ Any samples analyzed to verify that CCPs are under control. Analyses may involve physical, chemical, microbiological, or organoleptic methods

▲ Modifications to the HACCP plan

▲ Training and knowledge of individuals responsible for monitoring CCPs.

Figure 5.10—Guidelines for Determining How and When to Implement HACCP Plan Verification Procedures
(Source: *Food Code*)

Step 7—Establish an Effective Record Keeping System That Documents the HACCP System

An effective HACCP system requires the development and maintenance of a written HACCP plan. The plan should provide as much information as possible about the hazards associated with each individual food item or group of food items covered by the system. Clearly identify each CCP and the critical limits that have been set for each CCP. The procedures for monitoring critical control points and record maintance must also be contained in the establishment's HACCP plan.

The amount of record keeping required in a HACCP plan will vary depending on the type of food processing used from one food establishment to another. For example, a cook-chill operation in a campus dining hall would require more record keeping than a limited menu cook-serve operation in a neighborhood cafe. The detail of your HACCP plan will be determined by the complexity of your food production operation. **Keep sufficient records to prove your system is working effectively, but keep it as simple as possible**.

Changing a procedure at a CCP *but not recording the change on your flowchart almost guarantees that similar problems will repeat.* Therefore, record keeping is vital to the overall effectiveness of your HACCP system. A clipboard, work sheet, thermometer, watch or clock, and any other equipment needed to monitor and record these limits must be readily available to the food production staff. An example of what information to keep in records, as suggested by the *Food Code,* is presented in Figure 5.11.

---

1. List of HACCP team members and their assigned responsibilities

2. Description of the food product and its intended use

3. Flow diagram of the food preparation steps with CCPs noted

4. Hazards associated with each CCP and preventive measure

5. Critical limits

6. Monitoring systems

7. Corrective action plans for deviations from critical limits

8. Record keeping procedures

9. Procedures for verification of the HACCP system

Figure 5.11—Examples of Documents That Can Be Included in the Total HACCP System
(Source: *Food Code*)

The information presented in Figure 5.12 provides an example of how the information can be organized in your HACCP plan.

Steps in HACCP System   **Process Steps**

	Quick-Chill	Reheating
1. Hazard analysis	Bacteria (especially spore-formers)	Bacteria
2. Identify CCP	Yes	Yes
3. Identify Critical Limit	140°F to 70°F in 2 hours or less. 70°F to 41°F in 4 hours or less.	Reheat to an internal temperature of 165°F or above within 2 hours
4. Monitoring [Procedures Frequency Person(s) Responsible]	Cook monitors product temp. every 60 minutes	Cook monitors product temp. every 30 minutes
5. Corrective Action(s)/ Person(s) Responsible	Discard product if CCP limit is **not** met	Continue heating until CCP limit is achieved
6. Verification Procedure(s)/Person(s) Responsible	Food Manager	Food Manager
7. HACCP Records	Record data on a time/temp. chart and initial	Record data on a time/temp. chart and initial

Figure 5.12—Information Commonly Included in a HACCP Plan

The method you use to record the information is not especially important, as long as it is easy for the staff to use and provides quick access to the information contained in the log.

Education and Training

"Welcome to HACCP Training!"

Education and training are key to the successful implementation of a HACCP program. Your HACCP system should be integrated into each food worker's duties as well as performance plans and goals. Your employees' training program should be tailored to the operations they work.

The content of the training program should provide employees with an overview of the HACCP system and how it works to ensure food safety. The training program should also include an in-depth examination of the CCPs and critical limits in your HACCP system.

The primary goal of your HACCP training program is to make employees skilled in performing the specific tasks (monitoring and recording) which the HACCP plan requires them to perform. Motivate them by stressing the importance of their roles and their responsibility to the success of the program.

HACCP training must be an ongoing activity due to the high employee turnover that most establishments experience. Keep records of employee HACCP training along with the other documents generated by your HACCP system.

Effective training and supervision will help you achieve the benefits of a HACCP based operation in a much shorter period of time. This will save you money and, more important, enhance the safety of the products you are serving to your clients.

Roles and Responsibilities under HACCP

The role of health department personnel and other regulators is to promote the use of HACCP by the food industry. They can help you better understand the HACCP concept and how to implement a HACCP system. Regulatory personnel also will review your HACCP documents periodically to ensure that critical control points are properly identified, critical limits are properly set, required monitoring is being performed, and the HACCP plan is being revised when necessary.

The role of the food establishment managers and supervisors is to develop, implement, and maintain the HACCP system. You must continuously use and improve your HACCP system if safe food management is to be achieved.

Summary

If you implement a HACCP food safety system, you can avoid situations like the one described in the case at the beginning of this chapter. A HACCP recipe and flow chart would alert food workers to potential hazards and the critical control points that can be used to prevent, minimize, or eliminate them.

Poultry and poultry products have frequently been involved as the source of foodborne salmonellosis. If *Salmonella* bacteria are present after the cooking process, the bacteria can multiply to high levels that may not be destroyed later by the reheating processes. Poultry products should be cooked to an internal temperature of 165°F (74°C) (CCP-1) to kill the germs that may be present in and on the raw birds. The cooked turkey must be held at 140°F (60°C) or above (CCP-2). If not served immediately, the cooked food should be divided into smaller portions to ensure it is cooled to a product temperature of 41°F (5°C) within 4 hours (CCP-3). Temperatures should be checked periodically to ensure that cooling is occurring within this time frame. When reheating the turkey, food workers should be sure it reaches 165°F (74°C) in 2 hours or less (CCP-4). The reheated turkey should then be held at 140°F (60°C) or above until served (CCP-5).

A **hazard** is a biological, chemical, or physical property that may cause an unacceptable consumer health risk. If these hazards are not controlled, foodborne illnesses can occur.

The *Hazard Analysis Critical Control Point (HACCP)* system is a prevention-based safety program that identifies and monitors the hazards associated with food production. The HACCP system will help you estimate the relative risks for each hazard and determine where control procedures will be needed. The system, when properly applied, can be used to effectively control any area or point in the food flow that could cause a hazardous situation.

A **Critical Control Point (CCP)** is a step in the food flow where loss of control could result in an unacceptable risk to food safety. The most commonly used CCPs are cooking, cooling, reheating, and hot/cold holding. The *Food Code* also recognizes specific food handling and sanitation practices (such as proper thawing methods), cross contamination control, and certain aspects of employee and environmental hygiene as CCPs. The food industry considers these latter types of activities to be "standard operating procedures" or SOPs rather than CCPs. This is because handwashing and cross contamination control cannot always be accurately monitored and recorded. This should in no way diminish the importance of these activities and food safety practices. It merely acknowledges that they are harder to monitor and record than are critical control points.

Monitoring is a critical activity within a HACCP system. When done properly, monitoring can help pinpoint the cause of a problem and provide an early warning that a potentially hazardous situation might exist. If monitoring indicates that a critical limit has been exceeded, immediate action must be taken. Each food establishment is different and will have different CCPs. The corrective action must also be tailored to your particular establishment and the various food operations you use. Regardless of the type of operation, corrective actions must demonstrate that the CCP has been brought under control.

Food workers and managers share the responsibility for maintaining an effective record keeping system. Food workers should measure and record all appropriate monitoring data in HACCP records, and managers should review these records to make certain that monitoring is being done properly. Well-organized monitoring records provide evidence that food safety assurance is being accomplished according to HACCP principles.

The HACCP system is the most effective method created to date to ensure the safety of food processing and preparation operations. Implementation of a properly designed HACCP program will protect public health well beyond anything that could be accomplished using old style methods. Traditional inspections emphasized facility, equipment, design, and compliance with basic sanitation principles. HACCP focuses on the **actual** safety of the product.

Case Study 5.1

In the following exercise, a recipe for BBQ ribs is used to demonstrate application of the HACCP system. Use the recipe in Figure 5.13 and follow the decision tree in Appendix D to determine why each step is needed to ensure the safety of the finished food product.

Ingredients	Amount	Servings		
		25	50	100
Hoisin sauce	Tbsp.	2.5	5.0	10
Bean sauce	Tbsp.	2.5	5.0	10
Apple sauce	Tbsp.	2.5	5.0	10
Catsup	Tbsp.	5.0	10.0	20
Soy sauce	Tbsp.	2.5	5.0	10
Rice wine	Tbsp.	5.0	10.0	20
Peanut oil	Tbsp.	2.5	5.0	10
Ginger, minced	tsp.	1.25	2.5	5
Scallion, minced	1	1.5	5.0	10
Garlic cloves, minced	2	5.0	10.0	20
Sugar	Tbsp.	2.5	5.0	10
Salt	tsp.	1.25	2.5	5
Pork spareribs, trimmed	1 rack	2.6	5.0	10
Honey	Tbsp.	7.5	5.0	30

Figure 5.13—Sample Recipe and HACCP Flow Chart for Barbecued Spareribs

Pre-preparation

1. Thaw spareribs under refrigeration at 41°F(5°C) for 1 day or until completely thawed.

2. Wash hands before starting food preparation.

3. Combine all of the ingredients for the marinade in a large bowl.

Preparation

4. Score the ribs with a sharp knife. Place the ribs in the marinade. Marinate them for at least 4 hours at a product temperature of 41°F (5°C) or lower in refrigerated storage.

5. Place the spareribs in a smoker at 425°F (220°C) for 30 minutes.

6. Reduce the heat to 375°F (190°C) and continue smoking the ribs to a final internal temperature of 145°F (63°C) or higher. This phase of the cooking process should take about 50 minutes. Brush the honey on the spareribs during the last 5 minutes of the smoking process.

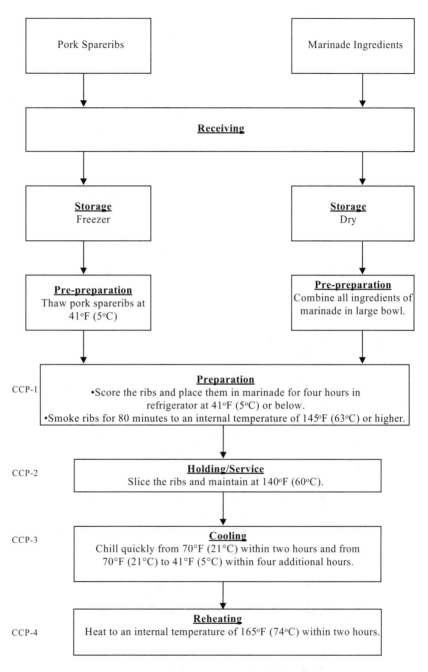

Figure 5.14—Flow Chart for Spareribs

(Adapted from: *HACCP Reference Book*, 1993.)

Holding/Service

7. Slice the ribs between the bones and maintain at 140°F (60°C) or above while serving.

Cooling

8. Transfer any unused product into clean, 2-inch deep pans. Quick-chill the product from 140°F (60°C) to 70°F (21°C) within 2 hours and from 70°F (21°C) to 41°F (5°C) within 4 hours.

Storage

9. Store the chilled product in a covered container that is properly dated and labeled. Refrigerate at 41°F (5°C) or below.

Reheating

10. Heat spareribs to an internal temperature of 165°F (74°C); or 190°F (88°C) in a microwave oven within 2 hours.

Case Study 5.2

An *E. coli* O157:H7 outbreak at a restaurant in Boise, Idaho, was traced to romaine lettuce used in salad. Thirteen confirmed and eight probable cases of the illness were reported during the outbreak. While one person who was ill required hospitalization, none experienced serious complications from their illness.

All of the victims had eaten at the same restaurant and 95% of them had eaten the same meal—a chicken Caesar salad. Any suspicions that the outbreak may have been caused by cross-contamination between beef and chicken were ruled out when it was learned that beef products were prepared on one side of the kitchen and chicken on the other. In addition, food workers were responsible for working with either beef or chicken but not both.

Romaine lettuce turned out to be the cause of the outbreak. Investigators tracked the lettuce back to its source and determined that contamination had occurred at the restaurant.

Sophisticated blood analysis showed that some food workers at the restaurant had been previously infected by *E. coli* O157:H7 bacteria. It was learned that one food worker had been shedding *E. coli* O157:H7 bacteria for several days.

▲ How could this situation have been prevented?

The answers to the case studies are provided in Appendix A.

Discussion Questions (Short Answer)

1. What is HACCP?

2. What are HACCP's advantages over traditional food safety programs?

3. What types of foods are most commonly incorporated into the HACCP system?

4. What is a hazard? List three classes of hazards and give an example of each one.

5. What is risk?

6. Define critical control point and give an example.

7. Why is ongoing monitoring important to a HACCP system?

Quiz 5.1 (Multiple Choice)

1. The Hazard Analysis Critical Control Point (HACCP) system should be employed:

   a. Whenever potentially hazardous foods are prepared, regardless of the type of food establishment.

   b. Only in institutional food facilities that provide food for very young or frail elderly consumers.

   c. Only in convenience stores where mechanical dishwashing equipment is not available.

   d. Only when foods are sold by a food establishment for consumption off site.

2. For which of the following products would it **not** be necessary to develop a HACCP flow chart?

   a. Chicken salad.

   b. Tuna salad.

   c. Seafood salad.

   d. Citrus fruit salad.

3. A **hazard** as used in connection with a HACCP system is:

   a. Any biological, chemical, or physical property that can cause an unacceptable risk.

   b. Any single step at which contamination could occur.

   c. An estimate of the likely occurrence of a hazard.

   d. A point at which loss of control may result in an unacceptable health risk.

4. A **risk** as used in connection with a HACCP system is:

   a. Failure to meet a required critical limit for a critical control point.

   b. An estimate of the likely occurrence of a hazard.

   c. Greatest for non-potentially hazardous foods.

   d. Lowest when complex recipes are required to produce a food item.

5. Which of the following statements about HACCP programs is **false**?

   a. The HACCP system attempts to anticipate problems before they happen and establish procedures to reduce the risk of foodborne illness.

   b. The HACCP system targets the production of potentially hazardous foods from start to finish.

   c. Records generated by the HACCP system can also be used to aid in foodborne disease investigations.

   d. A HACCP system should only be implemented by public health officials who have been certified by the FDA to conduct such programs.

6. Which of the following is **not** associated with the hazard analysis portion of the HACCP program?

   a. A review of the menu to identify potentially hazardous foods.

   b. A risk assessment of both the likelihood that hazards will occur and their severity if they do occur.

   c. A series of questions used to facilitate the identification of potential problems in each step in the flow of food.

   d. The identification of critical control points whereby hazards can be prevented, eliminated, or reduced to acceptable levels.

7. Which of the following statements is **false**?

   a. A critical limit is the threshold that must be met to ensure that each critical control point effectively controls a microbiological, chemical, or physical hazard.

   b. There must be at least two critical control points in the flow of food in order for a HACCP system to be implemented.

   c. Many steps in food production are considered control points, but only a few qualify as critical control points.

   d. A critical control point is a point, step, or procedure in food preparation where controls can be applied and a food safety hazard can be prevented, eliminated, or reduced to acceptable levels.

8. Which of the following statements about critical control point monitoring is **false**?

   a. Monitoring is a series of observations and measurements that is used to determine whether a critical control point is under control.

   b. The monitoring of critical limits can be performed either continuously or at predetermined intervals.

   c. Individuals who perform monitoring tasks must be management personnel who will do the job in an unbiased manner.

   d. An operation that identifies critical control points but does **not** establish a monitoring system, has not actually implemented a HACCP system.

9. The **ultimate** success of a HACCP program depends on:

   a. Eliminating potentially hazardous food items from your menu.

   b. Providing proper training and equipment for employees who are implementing the HACCP system.

   c. Having HACCP flow charts developed for all foods sold by the establishment.

   d. Food establishment managers having sole authority for implementing the HACCP system.

10. Which of the following is an example of a critical control point?

a. Poultry and eggs are purchased from approved sources.

b. Chicken and noodles are heated on the stove until the center of the product reaches 165°F (74°C) for 15 seconds.

c. Only pasteurized milk is used by the establishment.

d. The cutting board is washed and sanitized between chopping carrots and celery for the garden salad.

Answers to the multiple choice questions are provided in Appendix A.

References/Suggested Readings

Bryan, F. L. (1992). *Hazard Analysis Critical Control Point Evaluations: A Guide To Identifying Hazards and Assessing Risks Associated with Food Preparation and Storage.* World Health Organization. Geneva, Switzerland.

Federal Register, Volume 59, Number 149. U.S. Government Printing Office, 1994. Washington, DC.

Food and Drug Administration (1999). *1999 Food Code.* United States Public Health Service. Washington, DC.

Food Marketing Institute (1989). *A Program To Ensure Food Safety and Quality.* Washington, DC.

LaVella, B. and J. L. Bostic (1994). *HACCP for Food Service—Recipe Manual and Guide.* LaVella Food Specialists. St. Louis, MO.

J. K. Lotkin (1995). *The HACCP Food Safety Manual.* Wiley Press. New York, NY.

Pierson, M. D. and D. A. Corlett, Jr. (1992) *HACCP Principles and Applications.* Van Nostrand Rheinhold. New York, NY.

Rue, N., and National Assessment Institute (1994). *Handbook for Safe Food Service Management.* Prentice Hall. Englewood Cliffs, NJ.

The Educational Foundation of the National Restaurant Association (1993). *HACCP Reference Book.* National Restaurant Association. Chicago, IL.

Selected Web Sites

Garden State Laboratories
www.gslabs.com/haccp.html

Canadian Food Safety Sites Involving HACCP
foodnet.fic.ca/safety/safety.html

Iowa State University Extension
www.extension.iastate.edu/foodsafety/

International Food Information Council
ificinfo.health.org/

NSF International
www.nsf.org

International HACCP Alliance
ifse.tamu.edu/haccpall.html

National Restaurant Association
www.restaurant.org

The Food Marketing Institute
www.fmi.org

Gateway to Government Food Safety Information
www.foodsafety.gov

United States Department of Agriculture (USDA)
www.usda.gov

Centers for Disease Control and Prevention (CDC)
www.cdc.gov

United States Food and Drug Administration (FDA)
www.fda.gov

USDA Foods Safety and Inspection Service (FSIS)
www.fsis.usda.gov

USDA/FDA Food and Nutrition Information Center
www.nal.usda.gov/fnic/

Partnerships for Food Safety Education
www.fightbac.org

Food and Agriculture Organization
www.fao.org

6 FACILITIES, EQUIPMENT, AND UTENSILS

Lack of Proper Refrigeration Equipment Contributes to Armory Outbreak

Fifty-nine National Guardsman became suddenly ill following a holiday meal at the armory. Most of the victims experienced diarrhea, cramps, nausea, and vomiting.

None of the food or beverages served at the suspect meal were available for testing. However, the outbreak was likely caused by toxin produced by Staphylococcus aureus *bacteria. This conclusion is supported by the classic Staphylococcal food-poisoning symptoms of diarrhea, cramps, nausea and vomiting with a short incubation period. In addition, seven of the victims submitted stool samples to either their private physicians or to their admitting hospital laboratories for analysis. Three of the specimens were positive for toxin produced by* Staphylococcus aureus *bacteria.*

Staphlococcus aureus organisms are usually introduced into food when a food worker touches his or her nose, mouth, or an open sore and touches food without properly washing his or her hands first. Once present in the food, the Staphylococcus bacteria can multiply rapidly under optimal temperatures and produce a toxin in the food. The toxin is what causes the illness, and it is not destroyed when the food is cooked later.

During the foodborne disease outbreak, investigators suggested the refrigeration units at the Armory were not of adequate size to provide rapid cooling of potentially hazardous foods.

Learning Objectives

After reading this chapter, you should be able to:

▲ Identify ways in which the design and layout of facilities contribute to the efficiency and effectiveness of a food establishment

▲ Understand the importance of purchasing and properly maintaining equipment and utensils and the influence they have on food safety

▲ Describe how work tasks are conducted in work centers and how the preparation and service of food flows through a production area

▲ Identify the criteria that should be used when determining the need for each type of equipment

▲ Understand the basic design and construction requirements that apply to floor- and table-mounted equipment

▲ Explain the role of organizations such as NSF International and Underwriter's Laboratory Inc. in the approval of food equipment

▲ Recognize the different types of cooking, refrigeration, preparation and dishwashing equipment that are available for use in a food establishment

▲ Describe how proper installation and maintenance affect the operation of equipment used during the course of food production, storage, and service

▲ Explain the role of proper lighting in food production and warewashing

▲ Explain how proper heating, air conditioning, and ventilation affect food sanitation and worker comfort and productivity.

Essential Terms

Easily movable	Non-food contact surfaces
Easy-to-clean	Sealed
Equipment	Smooth
Food contact surfaces	Splash contact surfaces

Kitchenware
    Tableware
    Utensils
    Single-use articles	

Design, Layout, and Facilities

The design, layout, and facilities in a food establishment should be based on the menu being served and menu trends. The type of equipment used will be determined by the preparation procedures required to produce menu items. A good plan for one operation may not be good at another site with a different menu. Food facilities are normally used for long periods of time. Try to anticipate possible technological advancements and developments in foodservice equipment in the planning stages. Seek a design that facilitates future redesign and remodeling if needed to accommodate new menu items.

The design and layout of facilities, such as the kitchen, dishroom, and dining area, should provide an environment in which work may be done efficiently and effectively. The ideally designed facility enables the use of multiple sources of energy (electric and gas) as well as energy-efficient appliances and equipment.

Facilities planning and design are often left to architects or engineers who may not be familiar with the needs of the food establishment. You and your employees can help the architects or engineers by providing practical information about your operation before and during the design phase of the project. Use a team approach to prepare and review the facility plans. Try to avoid unnecessary expense and inconvenience caused by a faulty design.

The general areas of a food establishment are:

▲ Receiving and delivery

▲ Storage

▲ Preparation

▲ Holding

▲ Service

▲ Warewashing

▲ Garbage storage and pickup

▲ Food display area or dining room

▲ Housekeeping

▲ Toilet facilities.

Activities that are carried out in these different areas are called functions. Functions are further broken down into various sub-functions. When planning a food establishment, try to understand and visualize each and every function within that facility.

First, determine the tasks within the functions. Next, arrange these tasks in a way that allows a smooth and sequential flow within that area. It is important that you *develop a flow diagram* of your establishment's operation in order to plan the physical facilities for each function. A sample flow diagram is presented in Figure 6.1.

Attention to the design and construction of the building, the equipment, and its installation ensures high standards of sanitation and safety and helps make maintenance activities easier.

Regulatory Considerations

When planning facilities for food establishments, you must know of and comply with national, state, and local standards and codes, related to:

▲ Health

▲ Safety

▲ Building

▲ Fire

▲ Zoning

▲ Environmental code standards.

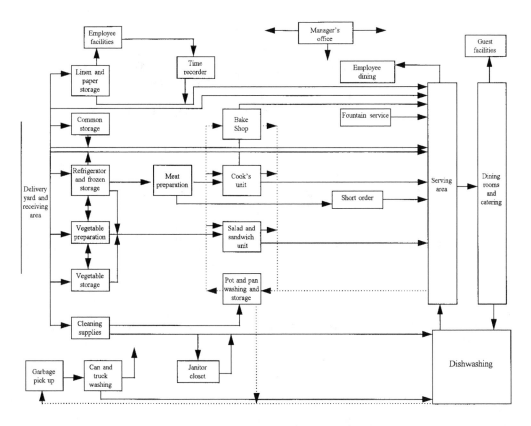

Figure 6.1—Flow Diagram Showing the Functional Relationships of Work

(Source: *Kotschevar and Terrell*, 1985)

The equipment used in food establishments should meet the standards and bear the stamp of approval of recognized standard organizations (Figure 6.2). This list includes:

▲ Nation Sanitation Foundation International (NSF)

▲ Underwriters Laboratories Inc. (UL)

▲ American Gas Association (AGA).

Buyers who purchase equipment approved by these organizations are assured that quality materials are used. Also, the items are built according to acceptable standards for the particular item. A list of NSF International Standards is provided at the end of this chapter. Installation of equipment must conform to local code. Officials from local health, building, and fire protection agencies usually conduct on-site inspections of equipment installation.

Figure 6.2—UL and NSF Seals

Work Center Planning

The food production area is commonly organized into **work centers**. These are areas where a group of closely related tasks are performed by an individual or individuals. The number of work centers required in a food facility depends on the number of functions to be performed and the volume of material handled.

Two of the most important features to consider when planning work centers are the total space needed and the arrangement of equipment in that space. A properly designed work center provides adequate facilities and space for:

▲ Efficient production

▲ Fast service

▲ A pleasant environment

▲ Effective cleanup.

A worker should be able to complete the related tasks at the work center without moving away from it. The work center should be large enough to do the job yet compact enough to reduce travel and conserve time and energy. A simply-arranged short-order work center showing the close relationship between production, dish-up, and service to dining room is presented in Figure 6.3.

Equipment Selection

It is extremely important to select the right piece of equipment for the job. Compare different pieces of equipment for a particular job and look at such features as:

▲ Design

▲ Construction

▲ Durability

▲ Ability to clean easily

▲ Size

▲ Cost

▲ Safety

▲ Overall ability to do the job.

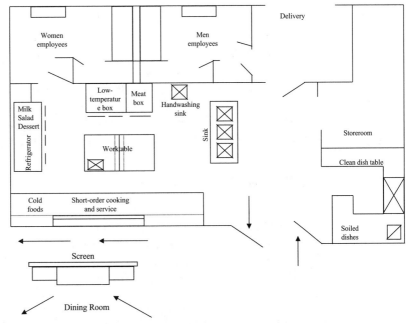

Figure 6.3—An Efficient Short-Order Cooking and Service Working Center

(Source: Kotschevar and Terrell, 1985)

Need

The basic needs of a food establishment should dictate the purchase of equipment. Need should be evaluated on the basis of whether the addition of a particular piece of equipment will improve the quality of food, reduce labor and material costs, improve sanitation, and contribute to the bottom line of the establishment.

Food establishment managers should assign a priority to their equipment needs. When buying equipment, look for products that meet current and future needs and demands of the operation, do

not require extraordinary repair and upkeep, and will work
properly for many years.

Cost

Some of the major costs associated with the purchase of equipment
are the purchase price, installation cost, operating costs,
maintenance costs, and finance charges.

Compare these costs for the different pieces of equipment that you
are considering before you make a purchase. Some of these costs
will be calculated by the manufacturers and provided for
comparison purposes. For others, the food establishment manager
or his agent will have to collect information to assist in making a
purchasing decision. It is not advisable to buy equipment that is
more expensive, larger, or more sophisticated than that which is
required to meet the needs of the operation.

Size and Design

Equipment should easily fit into the space available in the layout of
the facility. Consider future needs when selecting equipment, but it
is unwise to buy oversized equipment in anticipation of future
growth many years away. Oversized equipment can cause as much
wasted effort as equipment that is too small.

Design is also an important feature of food equipment, since the
equipment used in a food establishment is subject to constant use
and abuse. Equipment and utensils must be designed to function
properly when used for their intended purposes.

Equipment that sets on the floor and is not easily movable must be
elevated on legs or sealed to the floor. **When legs are used, they
should provide a minimum clearance of 6 inches (15 centimeters)
between the floor and the bottom surfaces of the equipment.** The
space makes it easier to clean the floor under and behind the
equipment and discourages pest harborage.

Heavy equipment such as ranges and cabinets may be mounted on
a raised masonry, tile, or metal platform and sealed to the floor.
This helps prevent trash and pests from collecting under the
equipment.

In a supermarket or convenience store, display shelving units,
display refrigeration units, and display freezer units located in the
consumer shopping areas do not have to be elevated or sealed to

the floor. However, the floor under these units must be maintained in a clean and sanitary manner.

Table mounted equipment (that is not easily movable) should be on 4 inch (10 centimeter) legs. This provides clearance between the counter top and the bottom of the equipment to make it easier to clean under and around the equipment.

Construction Materials

A piece of food equipment consists of food-contact and non food-contact surfaces. A food-contact surface is that part of the equipment with which food is likely to come into contact during production. Non food-contact surfaces are those parts of the equipment and the surrounding area with which food is not intended to come into contact with during production. However, non food-contact surfaces are frequently exposed to splash and spill during food production.

The *Food Code* and construction standards from NSF International and Underwriters Laboratories Inc. require food equipment and utensils to be:

▲ Smooth

▲ Seamless

▲ Easily cleanable

▲ Easy to take apart

▲ Easy to reassemble

▲ Equipped with rounded corners and edges.

Materials used in the construction of utensils and food-contact surfaces of equipment must be non-toxic and not impart colors, odors, or tastes to foods. Stainless steel is a preferred material for food contact surfaces.

Under normal use, the materials should be safe, durable, corrosion-resistant, and resistant to chipping, pitting, and deterioration.

Metals

We depend on metals for nearly everything in a food establishment. Chromium over steel gives an easily cleanable, high-luster finish. It is used for appliances, such as toasters and waffle irons, and trim where high luster is desired. Noncorrosive metals formed by the alloys of iron, nickel, and chromium may also be used in the construction of foodservice equipment.

Lead, brass, copper, cadmium, and galvanized metal must not be used as food-contact surfaces for equipment, utensils, and containers. This is because these metals can cause chemical poisoning if they come into contact with high acid foods (foods with a low pH).

Stainless Steel

Stainless steel is one of the most popular materials for food operations. It is commonly used for food containers, table tops, sinks, dish tables, dishwashers and ventilation hood systems.

The highly durable finish has a shiny surface which easily shows soil and makes it easier to clean and maintain. Stainless steel also resists rust and stain formation and can resist high temperatures.

Gauge refers to the thickness of sheet metal or the weight of the metal. As the number of the gauge increases, the thickness and weight of the metal decreases. Gauges between 8 and 22 are most commonly used in foodservice equipment.

Metal finishes are either polished or dull. However, the more polished the metal, the more easily it is scratched. Stainless steel, for example, must be polished and cleaned with great care.

Do not use abrasive cleaners and scouring pads to clean stainless steel because they scratch the surface of the metal which can become "germ farms."

The cost of stainless steel is based on the extent of polishing desired, because polishing requires labor, material, and energy. Finish No. 4 is most commonly preferred for food production areas whereas higher finishes are usually preferred in serving areas.

Plastic

Plastics and fiberglass are frequently used in foodservice equipment because they are durable, inexpensive, and can be molded into different combinations. Undoubtedly, plastics will be used more and more as components of food equipment in the future.

Some examples of plastics used in food establishments are:

▲ Acrylics (used to make covers for food containers)

▲ Melamines (used for a variety of dishes and glassware)

▲ Fiberglass (used in boxes, bus trays, and trays)

▲ Nylons (used in equipment with moving parts)

▲ Polyethylene (used in storage containers and bowls)

▲ Polypropylene (used for dishwashing racks).

Make sure you use only food-grade plastics. There are many different brand names for plastics. Select them on the basis of their use and durability. Harder, more durable plastics are easier to clean and sanitize. Always follow NSF Standard No. 51 when selecting plastic materials and coverings.

Wood

Wood has both advantages and disadvantages for use in foodservice equipment. It is light in weight and economical. However, the disadvantages outweigh advantages because of problems with sanitation. Wood is porous to bacteria and moisture, and it absorbs food odors and stains. Wood also wears easily under normal use, which requires frequent maintenance and replacement.

The *Food Code* permits the use of hard maple or an equally hard, close-grained wood for cutting boards, cutting blocks, and baker's tables. Wood is also approved for paddles used in candy kitchens and pizza establishments.

Figure 6.4—Wood vs. Plastic Cutting Boards

Types of Equipment

In order to select the proper equipment for a food establishment, it is necessary to obtain up-to-date information about the equipment from manufacturers' brochures and catalogues, and trade journals and magazines.

NSF International has developed uniform equipment construction standards and provides independent laboratory testing of equipment and materials. The NSF seal on a piece of equipment indicates that it meets the requirements set in the standards. Sample pieces of the same design and model have been tested for performance in the laboratory and in the field. For additional information refer to the reference/suggested reading section at the end of this chapter.

Cooking Equipment

The types and quantities of food to be prepared in a food operation are prime considerations in selecting cooking equipment. Always consider durability and energy conservation.

The frame, door, exterior, and interior materials should not take away from the durability and ability to clean the unit as a whole. Insulation materials and their thickness contribute to the energy efficiency of this type of equipment.

For the sake of time and convenience, only buy equipment that has been tested and approved by such organizations as the American Gas Association, Underwriters Laboratories (UL), and NSF International.

Ovens

Ovens are among the most important pieces of equipment in a food establishment. They are used to cook a variety of foods to different temperatures. The heat in an oven is distributed by radiation, conduction, or convection, depending on the type of oven being used.

A good oven should rise to 450°F (232°C) within 20 minutes, and proper heat circulation is important. Ovens should be able to cool quickly when a drop in temperature is required.

The *range oven* is the most commonly used oven in small food operations (Figure 6.5). It has a cooking surface above the oven which makes it suitable for doing all kinds of work. Range ovens

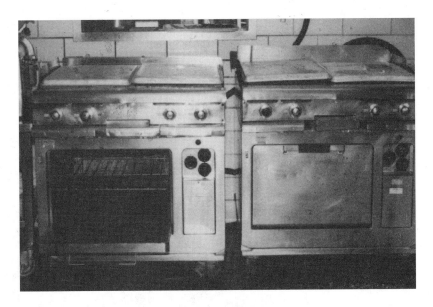

Figure 6.5—Range Oven

must have good insulation and construction to prevent heat loss. They should also be equipped with heavy-duty doors and shelves. A window in the oven door makes it easier to see inside.

Deck ovens are the stationary type where one deck sits on top of another to conserve space (Figure 6.6). A pizza oven is a common example of a deck oven.

In these units, the heated air is circulated around the outside of the heating chamber and radiates through the lining. Deck ovens contain separate heating elements and controls for each unit, and good insulation is provided between decks. The decks are designed to provide different working heights with 7-8" clearance provided for baking and 12-16" clearance for roasting.

A ***convection oven*** uses a high-speed fan to circulate heated air and guarantee even-heat distribution over and around foods (Figure 6.7). Cooking time in a convection oven is about 1/3 to 3/4 the amount required in deck ovens. Another advantage of the convection oven is that more cooking is accomplished in a smaller space, as food is placed on multiple racks instead of on a single deck.

Courtesy of Hobart Corp.

Figure 6.6—Deck Oven

Microwave ovens are commonly used for thawing, heating, and reheating food (Figure 6.8). Microwaves are electromagnetic waves of radiant energy which, like light, travel in straight lines. Microwaves are reflected by metals, pass through air, and are absorbed by several food components. They pass through many, but not all, types of paper, glass, and plastic materials. When microwaves are reflected or pass through a material, they do not give off heat to the object. To the extent that microwaves are absorbed, they heat the absorbing material.

The advantage of the microwave oven is that it can cook small quantities of food very rapidly. A microwave oven is not useful for cooking large quantities of food due to its limited capacity. However, it is excellent as a support to quantity cooking.

Many other types of ovens are available for use in a food establishment. These include conveyor, roll-in, revolving, and infrared ovens. Each of these pieces of equipment has unique features and has been designed for special applications. You should contact your equipment supplier to determine if this type of equipment is appropriate for your operation.

Courtesy of Hobart Corp.

Figure 6.7—Convection Oven

Figure 6.8—Microwave Oven

Steam-jacketed Kettles

Steam-jacketed kettles are essential pieces of equipment in most food establishments. These kettles consist of two bowl-like sections of welded aluminum or stainless steel with an air space between for circulation of steam (Figure 6.9). When the steam is released inside the jacket, it condenses on the outside of the inner shell, thereby giving up its heat to the metal, from which it is then transferred to the food. The steam does not come directly in contact with the food being heated.

Steam-jacket kettles may be stationary or tilting. The cooking capacity of kettles should be based on a maximum load to fill the kettle to about 2/3 to 3/4 full. More than that makes it difficult to handle and pour the food without spillage. When planning for kettles, it is important to remember that emptying the contents of the container should be as easy as possible.

Courtesy of Groen Corporation

Figure 6.9—Steam Kettle

Refrigeration and Low-Temperature Storage Equipment

Refrigeration and low-temperature storage equipment are widely used to protect perishable foods and preserve wholesomeness.

Many foods deteriorate rapidly at room temperature. A lot of waste can be eliminated by keeping foods at lower temperatures until they are used.

Proper cooling requires removing heat from food quickly enough to prevent microbial growth. Improper cooling of potentially hazardous foods is consistently identified as one of the leading contributors to foodborne illness. Bacteria grow best at temperatures between 70°F (21°C) and 120°F (49°C). **The *Food Code* states that cooked potentially hazardous foods *must* be cooled from 140°F (60°C) to 70°F (21°C) within 2 hours and from 70°F (21°C) to 41°F (5°C) or below within 4 hours.**

Other potentially hazardous foods, such as reconstituted foods and canned tuna, must be cooled to 41°F (5°C) or below within 4 hours if prepared from ingredients that are at room temperature. A potentially hazardous food, such as shell eggs, received at a temperature *above* 41°F (5°C) during shipment from the supplier, must be cooled to 41°F (5°C) or *below* within 4 hours.

Refrigeration is important during the transportation and storage of perishable foods. Together they allow us to have an ample supply of meats, poultry, fish, dairy products, fruits, and vegetables in all parts of the country practically all seasons of the year. Refrigeration during transportation and better cold storage facilities have helped to stabilize the price of perishable foods.

The efficient operation of a refrigeration unit depends on several factors including:

▲ Design

▲ Construction

▲ Capacity of the equipment.

Proper air circulation both inside and outside the cabinet is important. **The storage of food in shallow containers, placed on slatted shelves or tray slides to permit good circulation of the chilled air**, is essential for both short- and long-term storage.

Sheet pans, foil, plastic, or cardboard should not be used to line shelves because they decrease air flow. Sturdy construction of doors, hardware, and fixtures is a necessary requirement. Doors may be full or half length, and shelving should be adjustable. Door gaskets should provide reasonable wear and be replaceable when worn.

The size of refrigerator or freezer needed depends on the size of the work area and the type and amount of food to be stored in the unit. Refrigerators and freezer units have far more capacity today than ever before. More compact compressors are usually located at the top of the unit, which means the unit can be taller, utilizing all the space and increasing capacity. The type and size of refrigeration system selected should insure reliable and efficient operation.

All types of refrigerators and freezers have a maximum capacity for cooling foods. Too often, workers put large amounts of hot food in a unit and rely on it to "cool" the food. However, the *addition of large amounts of hot food will cause the inside temperature of the unit to rise above acceptable storage temperatures.* This, in turn, could cause food to be stored in the temperature danger zone. Do not forget that the primary purpose of a refrigerator is to hold foods that are at or near the proper cold storage temperature. Cold foods must not be stored above the maximum load line in a chest type refrigerated display case (Figure 6.10).

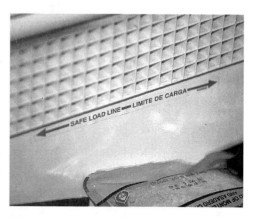

Figure 6.10—Maximum Load Line in Refrigerated Display Case

Maintenance of the refrigerated equipment in the kitchen and other areas is an important responsibility. Clean refrigerators on a regular basis to maintain good sanitary conditions and eliminate odors. Wash the inside walls, floor, shelves and other accessories of a walk-in unit as needed to clean up spills and remove debris. Clean the fan grates and condenser unit periodically.

Open doors of refrigerators as briefly and infrequently as possible when taking food in and out of storage. Large amounts of warm, moist air raise the holding temperature and humidity inside a refrigerator.

Reach-in Refrigeration

There are several types of reach-in refrigeration units on the market. These include *pass-through, upright, under-the-counter, display cabinets, mobile* or *portable*, and *dispensing units*, to name a few.

In a small operation, reach-in refrigerators may provide all of the cold storage that is needed. In larger operations, reach-ins may serve the storage needs of a single work center.

The capacity of reach-in refrigerators varies greatly. Small units will have only a few cubic feet of storage space while large models have a capacity of about 90 cubic feet. Upright models are generally 6 feet tall, 3 feet deep, and come in various widths. These units are usually equipped with single, double, or multiple doors [Figure 6.11 (a)].

a. Standard Upright   **b. Under-the-Counter**

Courtesy of Hobart Corp.

Figure 6.11—Reach-in Refrigerators

Pass-through refrigerator units open from both sides and are used for storage between production and service areas. Pass-through refrigerators save labor and eliminate interference by providing access of two working areas simultaneously.

However, the ability to open these units from two sides may increase the loss of cold air if the doors are opened repeatedly or for long periods of time. Keep the doors to these refrigerators closed tight, except when food is being put into or taken out of the unit.

Another example of an in-line production refrigerator is the under-the-counter drawer unit, placed below the counter or under equipment [Figure 6.11 (b)]. It provides instant, on-the-spot storage. These units can be placed below fryers, griddles, and salad preparation areas.

Walk-in Refrigerators

Walk-in refrigerators are used to store large quantities of food in bulk at temperatures between 32° and 41°F (0° and 5°C). They are also used to safely thaw products removed from a freezer. Locate walk-ins to make the delivery of foods from receiving to production and service areas as easy as possible. The floors of walk-ins should be made of strong, durable, easily cleanable material. The walls need a washable finish that will resist the effects of moisture. A safety device for opening the door should be provided on the inside of the unit.

Some walk-in units have the door opening covered with four-inch wide plastic strips called a strip curtain. These curtains reduce air loss and conserve energy when the door is opened. An outside wall mounted thermometer or recorder will enable you to determine the air temperature inside the walk-in cooler without opening the door. The sensing part of the thermometer must be placed in the warmest part of the cooler.

Cook-Chill and Rapid-Chill Systems

Cook-chill is a system in which food is:

▲ Cooked using conventional cooking methods
▲ Rapidly chilled using a chiller
▲ Stored for a limited time
▲ Reheated before service to the customer.

In a standard cook-chill system, the food is cooked and then placed in sealed plastic pouches. The pouches are placed in a chill tumbler which looks similar to a front loading clothes washing machine. The pouches are then tumbled inside the machine in ice cold water. The food is typically chilled to 37°F (4°C) in 90 minutes or less and is stored at temperatures between 33°F and 38°F (1° and 4°C) for five days. Day one is counted as the day of production and day five is the day of service.

The advantages of this system include reduction of peaks and valleys in production and readily available foods. The disadvantages of this system include the high cost of the chiller and the storage-space requirements.

Rapid-chill systems are designed to cool hot foods very quickly. This type of equipment can typically get a few hundred pounds of hot food through the temperature danger zone in two hours or less. Although this equipment is somewhat expensive, it can be a very good investment for many food establishments.

Courtesy of Hobart Corp.

Figure 6.12—Rapid-Chill Equipment

Other Types of Food Equipment

Slicers

The basic design of food slicers includes a circular knife blade and carriage that passes under the blade (Figure 6.13). Foods to be

Courtesy of Hobart Corp.

Figure 6.13—Slicer

sliced are placed on the carriage and fed either automatically or by hand. Thickness of the slices can be controlled by changing the setting of a close adjustment to the blade.

Slicers can be dangerous if not used properly. The manufacturer's instructions should always be followed and employees should receive training on safe operation and proper cleaning of equipment.

Mixers, Grinders, and Choppers

Mixers, grinders, and choppers are available as table and floor models (Figure 6.14). Mixers can also shred and grind by using different accessories and attachments.

Floor model mixers have three standard attachments: (1) a paddle beater for general mixing which can be used to mash, mix, or blend foods and ingredients; (2) a whip to incorporate air into products; and (3) a dough hook used to mix and knead dough.

Grinders and choppers work well with a variety of fresh foods and ingredients including meats and vegetables.

Courtesy of Hobart Corp.

Figure 6.14—Meat Chopper

Ice Machines

Ice is food, and it must be handled with the same degree of care as other foods. Ice must be made from potable water, and ice machines must protect the ice during production and storage (Figure 6.15).

The parts of an ice machine that contact the ice are considered food contact surfaces. These surfaces must be smooth, durable, easily cleanable and constructed of non-toxic materials. Food contact surfaces of ice machines must be cleaned and sanitized regularly to prevent the growth of mold and other microorganisms.

The drain line from the ice machine must be equipped with an air gap to prevent the backflow of contamination into the ice. You will learn more about these concepts in Chapter 8.

Scoops used to dispense ice must be designed and constructed to meet the basic requirements of food equipment. Scoops must be stored either outside the ice machine in a protected fashion or inside the machine in a bracket that is mounted to a wall of the ice

Figure 6.15—Ice Machine

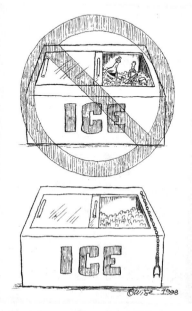

Figure 6.16—Do Not Store Food and Beverage Containers in

Ice Served to Customers

storage compartment. In either instance, the scoop must be stored in such a way that the handle can be grasped by workers without their hands coming into contact with the ice. Workers must never dispense ice by passing a glass or cup through the ice. This may cause the glass from the container to break off and become a physical hazard in the ice. Food and beverage containers should not be stored in ice that will be used for drinking purpose. This helps prevent contamination of the ice that your customers may consume.

Dishwashing Equipment

The effective cleaning and sanitizing of equipment and utensils is one of the most important jobs in a food establishment. Contaminated utensils, equipment (e.g., slicers), and cutting boards have been identified as sources of cross contamination of food during preparation and service. Contaminated eating utensils can aid the spread of bacteria and viruses that can cause disease.

Dishwashing is best performed in a separate room or area which is well lighted and ventilated. Noise-absorbing materials are used to aid in lowering the noise levels associated with this activity. The design of the dishwashing area is dependent upon the total volume of equipment and utensils to be washed and the time required to do the job. As business changes, periodically evaluate the dishwashing operation to make certain that it can effectively handle the current volume.

Food and beverage containers should **not** be stored in ice that will be used for drinking purposes. This helps prevent contamination of the ice that goes into your customer's glass.

The items that are most frequently washed and sanitized in a food establishment are kitchenware and tableware. Kitchenware includes kitchen utensils such as pots, pans, and skillets soiled in the process of food preparation. Tableware includes dishes, glassware, and eating utensils that are handled by the employees or customers.

The purpose of dishwashing is to clean and to sanitize equipment, dishware, and utensils. It requires a two-part operation:

▲ **A cleaning procedure to remove visible soil from the surface of the item**

▲ **A sanitizing procedure to reduce the number of disease-causing microorganisms on a cleaned surface to safe levels.**

Cleaning and sanitizing kitchen equipment and utensils can be done manually or mechanically. Manual cleaning and sanitizing is typically used for in-place cleaning and sanitizing of equipment and for cleaning and sanitizing portable equipment and utensils. Most food establishments will use manual dishwashing to clean and sanitize pots, pans, and utensils used to prepare and serve food.

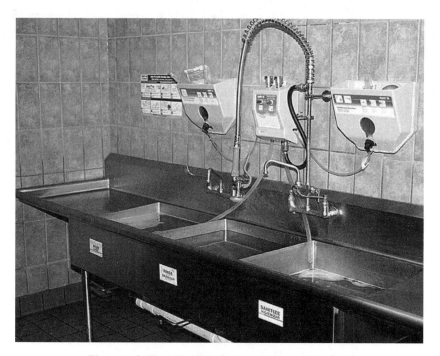

Figure 6.17—Three-Compartment Sink

Manual Washing

Sufficient space to store soiled equipment and utensils must be provided in the manual dishwashing area. Equipment and utensils must be pre-flushed or prescraped and, when necessary, presoaked to remove food particles and soil. Provide a hose and nozzle or similar device for pre-flushing and pre-scraping food soil into a garbage container or disposal. Locate this equipment at the soiled end of the manual dishwashing operation to prevent the contamination of cleaned and sanitized equipment and utensils.

Use a three-compartment sink for manual cleaning and sanitizing (Figure 6.17). In the first compartment, wash the scraped equipment and utensils in a detergent solution with hot water.

Rinse the soap off the equipment and utensils in the second compartment. Use the third compartment to sanitize. There are two accepted methods for sanitizing equipment and utensils:

▲ The hot-water method
▲ The chemical method.

Refer to Chapter 7 of this text for more information about proper cleaning and sanitizing techniques.

The compartments of the three-compartment sink must be large enough to accommodate the largest pieces of equipment and utensils used in the food establishment. Supply each compartment with hot and cold potable running water. Provide drainboards or easily movable dish tables of adequate size, for proper handling of soiled utensils prior to washing, and for air drying cleaned and sanitized items. **In a manual operation, the temperature of hot water used for sanitizing must be maintained at or above 171°F (77°C).** To achieve water temperature that high, you will need a booster heater or an electrical heating element that can be immersed directly in the water in the third compartment.

Mechanical Washing

Mechanical dishwashers are among the most expensive and important pieces of equipment in a food establishment. Use them for cleaning and sanitizing any equipment or utensil that does not have electrical parts and will fit into the machine. The most common types of mechanical dishwashers are:

Immersion dishwashers, where racks of dishes are immersed for cleaning. This arrangement is very similar to the Manual dishwashing procedure already discussed. *Single-tank, stationary-rack dishwashers*, where dishes are placed on racks and washed one rack at a time with jets of water within a single tank. They are operated by opening a door, inserting a rack of dishes, closing the door, and starting the machine (Figure 6.18).

Conveyor-rack dishwashers, where a conveyor carries the racks of dishes through the dishwasher. These machines are built with either single or multiple tanks. The racks move through the machine automatically. to the manual dishwashing procedure already discussed.

Flight-type dishwashers, where the conveyor is one continuous rack and the dishes are placed on pegs or bars. This machine operates continuously with a person stationed at each end (Figure 6.19).

Courtesy of Hobart Corp.

Figure 6.18—Single Tank Dishwashing Machine

Carousel-type dishwashers, where there is a closed circuit conveyor where dishes are loaded and unloaded.

Low-temperature dishwashers, which use chemicals to sanitize equipment and utensils. This allows lower water temperatures, which conserves energy.

The design of the dishwasher should fit in the overall layout of the food operation. The capacity needed will depend on the number of dishes to be cleaned. Specifications are provided by manufacturers for different types of dishwashers. When making a selection, you should also consider the cost of installation and maintenance.

An adequate supply of very hot water (180° to 195°F, 82° to 91° C) is required for the final rinse in a high-temperature dishwashing machine. The extremely high temperatures required for effective sanitization make it expensive and unsafe to maintain general-purpose hot water at these temperatures. Therefore, a separate booster heater is needed to raise the temperature of the sanitizing rinse water to between 180° and 195°F (82° and 91° C). The booster heater must be properly sized, installed, and operated to deliver water at the volume, flow pressure, and temperature required for the operation. For more information about booster heaters, consult NSF International Standard No. 5, *Commercial Hot Water Generating Equipment.*

Figure 6.19—Flight type Dishwasher Machine
Courtesy of Hobart Corp.

Installation

Proper installation is a necessity for the successful operation of all equipment. The best design and construction would be worthless if electrical, gas, water, or drain connections were inadequate or poorly done. The dealer from whom the equipment was purchased may or may not be responsible for its installation. Arrangements for installation will usually be specified in your purchase agreement.

Architects, engineers, and contractors are responsible for providing:

▲ Adequate plumbing

▲ Electrical wiring

▲ Venting facilities for the satisfactory installation of kitchen equipment

▲ Compliance with the standards of local building, plumbing, electrical, health, and fire safety codes.

The operation of each piece of equipment must be checked many times by both the contractors and service engineers before it is ready for actual use. Provide full instruction for the proper operation and satisfactory performance of each piece of equipment to all persons who will work with it. They must know the danger signals so that preventive measures can be taken early.

Maintenance and Replacement

The cost of care and upkeep on a piece of equipment may determine whether or not its purchase and use are justified. Successful maintenance of equipment requires definite plans to prolong its life and maintain its usefulness. Such plans place emphasis on a few simple procedures:

▲ Keep the equipment clean

▲ Follow the manufacturer's printed directions for care and operation

▲ Post the instruction card for a piece of equipment near it

▲ Stress careful handling as essential to continued use

▲ Make needed repairs promptly.

Some valuable suggestions for the care of equipment are:

▲ Assign the care of a machine to a responsible person

▲ Check the cleanliness of machines daily

▲ Have repairs performed promptly and by a competent person.

Lighting

Good lighting is a very important feature of a work area. Proper lighting in the kitchen increases productivity, improves workmanship, reduces eye fatigue and worker irritability, and decreases accidents and waste due to worker error. Proper lighting also shows soil and when a surface has been cleaned.

The kitchen should be equipped with proper lighting and soft colors which reduce glare. Locate light to eliminate shadows on work surfaces or glare and excessive brightness in the field of vision. Glaring light is unacceptable because it tends to harm vision, cause fatigue, reduce efficiency, and increase accidents.

The proper intensity of light depends on the kind of work to be done. The following intensity levels are recommended in the *Food Code*:

▲ *10 footcandles at a distance of 30" above the floor*
   ▼ Walk-in refrigeration areas
   ▼ Dry food storage areas
   ▼ Other areas and rooms during periods of cleaning.

▲ *20 footcandles at a distance of 30" above the floor*
   ▼ Areas where fresh produce or packaged foods are sold or offered for consumption
   ▼ Areas used for handwashing
   ▼ Areas used for warewashing
   ▼ Areas used for equipment and utensil storage
   ▼ Toilet rooms.

▲ *50 footcandles at the work surface*
   ▼ Where a food employee is working with unpackaged, potentially hazardous food
   ▼ Where a food employee is working with food, utensils, and equipment such as knives, slicers, grinders, or saws
   ▼ Where employee safety is a factor.

The recommended intensity of light for most dining areas is 15 to 20 footcandles. The food regulations in some states and local jurisdictions may require higher illumination levels in food establishments, especially during periods of cleaning. Consult the food regulation in your jurisdiction to determine what lighting levels are required for your establishment.

Light bulbs must be shielded, coated or otherwise shatter-resistant when used in areas where there is exposed food; clean equipment, utensils, and linens; or unwrapped single-service and single-use utensils (Figure 6.20). This is to prevent glass fragments from getting into food and onto food contact surfaces. Shielded or shatter-resistant bulbs are not required in areas where packaged foods are stored provided the packages will not be affected by broken glass falling onto them and the packages are capable of being cleaned of debris from broken bulbs before the packages are opened.

Light bulbs over self-service buffets and salad bars must also be shielded or shatter-resistant. In addition, this lighting must not mislead customers about the quality of the food items that are on display.

a. Glass and food don't mix

b. Plastic coated shatterproof bulbs

c. Shielded lighting over exposed foods

Figure 6.20—Protect food, equipment, linens, and unwrapped single-use utensils from broken glass

(a) and (b) used with permission of Shat-R-Shield, Inc.

Heating, Ventilation, and Air Conditioning (HVAC)

An adequate supply of fresh air, suitable temperature, and humidity are necessary for worker comfort and productivity. A worker's comfort and well-being also require air that is free of obnoxious odors and other objectionable or injurious elements. Achieving this in food establishments where heat, odors, and high air moisture are commonly created presents a challenge.

Air conditioning in food establishments means more than simply "cooling the air." It includes heating, humidity control, circulation, filtering, and cooling of the air. Modern HVAC systems will filter, warm, humidify, and circulate the air in the winter and maintain a desirable and comfortable temperature in the summer.

Kitchen ventilation is typically provided by means of mechanical exhaust hood systems. These systems keep rooms free of excessive heat, steam, condensation, vapors, obnoxious odors, smoke, and fumes. The standard ventilation system used in food establishments consists of a hood, fan, and intake and exhaust air ducts and vents (Figure 6.21). Ventilation hood systems must be designed and constructed to prevent grease or condensation from dripping back onto food, equipment, utensils, linens, or single-service and single-use articles. The capacity of the ventilation system should be based on the quantity of vapor and hot air to be removed.

Figure 6.21—Ventilation Hood System

Hoods are usually constructed of stainless steel or a comparable material that provides a durable, smooth, and easily cleanable surface. The hood should be equipped with filters or other grease-extracting equipment to prevent drippage onto food.

Filters or other grease-extracting equipment must be designed to be removed easily for cleaning and replacement or designed to be cleaned in place. Many fires at food establishments are caused by grease accumulation in filters and ducts. Filters should be cleaned or replaced regularly to ensure that they are working efficiently. This must be a part of your regular cleaning program.

Schedule the frequency for filter cleaning or replacement on the amount and type of pollution products in the air exhausted from the kitchen. The National Fire Protection Association recommends that all hoods be equipped with a fire extinguishing system.

Clean intake and exhaust air ducts so they are not a source of contamination of dust, dirt, and other materials. If vented to the outside, ventilation systems must not create a public health nuisance or unlawful discharge.

Pollution control devices that can be placed in hoods and ventilators are available. These devices operate with both hot water spray from nozzles, and electrostatic precipitators which use differences in electrical charge to remove particles.

Summary

Investigators are not exactly sure how the *Stahylococcus* bacteria identified in the case study at the beginning of the chapter was introduced into the food. However, they suspect that the food was maintained at an improper temperature, which promoted the multiplication of the bacteria and toxin production. If not served immediately, cooked food should be rapidly cooled to a temperature of 41°F (5°C) within 4 hours in order to limit bacterial growth. Investigators recommended that the Armory replace the current refrigerators with larger units or use alternative methods of rapid cooling such as ice baths or storing food in shallow pans rather than in large quantities.

Good planning is essential for maximum productivity and efficiency. The total alloted space needed and the arrangement of equipment in the space are the overall features you must consider when planning the layout of a food operation. The design of your facility should encourage a smooth flow of work, provide adequate facilities and space for completing the work efficiently, and contribute to high sanitation standards.

The basic needs of a food establishment should determine the type and size of equipment purchased. When purchasing equipment, look for items that can meet the current and future needs of the operation, can be properly cleaned and sanitized, and do not require extraordinary repair and upkeep.

The size and amount of food equipment and placement in the facility should be determined by managers of the food establishment working closely with an experienced engineer or architect. Ideally, equipment should easily fit into the available space.

Materials used in the construction of food contact surfaces should be safe, durable, corrosion-resistant, and nonabsorbent. These materials must not allow the transfer of harmful substances, or allow the transfer of colors, odors, or tastes to food.

The equipment used in food establishments should meet the standards and bear the seal of approval of the National Sanitation Foundation, Underwriters Laboratories Inc., or other recognized standards organizations.

Always install equipment in accordance with local building, plumbing, electrical, health, and fire safety codes. The best possible design and construction of equipment is worthless if electrical, gas, water, or drain connections were inadequate or poorly installed.

Case Study 6.1

The O.K. Corral Restaurant is a buffet style operation that offers customers a choice of roast beef, ham, or turkey in addition to several hot vegetables. The meat and poultry on the buffet are placed on wooden cutting boards and are carved in the dining room in front of the customers. The temperature of the meat and poultry is out of the danger zone when it leaves the kitchen. However, no effort is made to keep these products hot once they reach the dining room. The carver periodically wipes his carving utensils with a damp kitchen towel.

1. What food safety risks exist at the O.K. Corral?

2. What steps should the manager take to correct the food safety problems that exist in the facility?

Case Study 6.2

Mark Sellmore is the deli manager at Longfellow's Supermarket. Mark has been given permission to purchase a new slicer for the department. Mark's store manager has given him only one instruction. "Buy the most equipment possible for the money you spend."

1. What strategy should Mark use when purchasing the new slicer?

2. What factors should Mark use to determine the deli's *need* for the slicer?

3. What design and construction features should Mark use when he is making comparisons between slicers?

4. What standards organizations might Mark consult when evaluating different kinds of slicers?

Case Study 6.3

A gastrointestinal disease outbreak affected 123 of the 1,200 individuals who attended a meeting at the convention center in Columbus, Ohio. The initial assumption that the outbreak was foodborne did not pan out. Though two food items were significantly associated with the illness, their connection proved to be coincidental.

Based on the incubation period, symptoms, and duration of the illness, investigators speculate that *Giardia lamblia* and a Norwalk-like virus caused the illness. Because both agents are commonly waterborne, investigators turned their attention to the water supply at the convention center. Of particular interest was an ice machine which supplied the portable bars. It was discovered that a flexible tube from the machine had been inserted into a floor drain. When the machine's filters were serviced, the tube formed a connection between the ice machine's water supply and raw sewage.

▲ What type of plumbing hazard contributed to the foodborne disease outbreak discussed in the case study?

The answers to the case studies are provided in Appendix A.

Discussion Questions (Short Answer)

1. In what ways do layout, design, and facilities planning influence a food establishment?

2. What is a work center? What layout features are important to the efficiency of a work center?

3. What design and construction factors should a manager consider before buying a particular piece of equipment?

4. What standards organizations are commonly involved in the testing of food equipment? How can these organizations be helpful to a food establishment manager?

5. What kinds of materials are commonly used in the construction of equipment and utensils? What are the advantages and disadvantages of each kind of material?

6. How does a food contact surface differ from a non-food contact surface?

7. What are some features that make a piece of equipment easily cleanable?

8. How do proper lighting and ventilation affect sanitation in a food establishment?

Quiz 6.1 (Multiple Choice)

1. The design, layout and facilities in any food establishment should be based on the:

   a. Number of employees working in the kitchen.
   b. Number of meals served each day.
   c. Menu to be served.
   d. Amount of space available in the kitchen.

2. Which of the following organizations does **not** routinely evaluate the design and construction of food equipment?

   a. The United States Food and Drug Administration.
   b. NSF International.
   c. The American Gas Association.
   d. Underwriters Laboratories Inc.

3. Which of the following criteria is least important when determining a food operation's need for a particular piece of equipment? Will the equipment:

   a. Result in improved quality of food?

   b. Produce a significant savings in labor and material?

   c. Improve sanitation?

   d. Make the facility more attractive to customers?

4. When buying equipment, food establishment managers should look for equipment that:

   a. Can meet the need and demands of the operation.

   b. Can be used for a long time.

   c. Does not require excessive repair and upkeep.

   d. All of the above.

5. Which of the following statements is **false**?

   a. The size of equipment should be such that it can easily be fitted in the space available in the layout of the facility.

   b. Construction requirements for equipment vary according to whether the surface or area is a "food contact surface" or a "splash zone."

   c. Wood is a preferred material for food contact surfaces of equipment because it resists moisture and bacterial growth.

   d. Floor-mounted equipment that is not easily movable must be elevated on 6-inch legs or sealed to the floor on a base that is sealed to the floor.

6. Which of the following statements about ovens is **false**?

   a. A convection oven cooks food by circulating heated air around the product.

   b. Newer microwave ovens are sized to permit quantity cooking in a food establishment.

   c. The range oven is the type of oven most commonly used in food establishments.

   d. Deck ovens contain separate heating elements and controls for each unit.

7. Which of the following statements about refrigeration equipment is **false**?

    a. The size of refrigerator or freezer needed depends on whether a walk-in unit is available, what is going to be stored, and how much.

    b. Mechanical refrigeration during transportation and storage is important to preserve perishable foods.

    c. Pass-through refrigeration units open from both sides and are used for storage between production and service areas.

    d. The cleanliness and care of refrigerated equipment is not important, since it is only used to store packaged foods.

8. What type of equipment is required when a food establishment uses a manual dishwashing operation?

    a. A three- or four-compartment sink with hot and cold potable running water and drainboards or easily movable dish tables.

    b. A two-compartment sink with hot and cold potable running water and drainboards or easily movable dish tables.

    c. A two-compartment sink with suitable facilities for pre-flushing and prescraping the dishes before they are washed.

    d. A single-tank dishwashing machine that is large enough to accommodate the largest pieces of equipment and utensils used in the operation.

9. From a sanitation perspective, what is the **most** important reason for having adequate lighting in a food production area?

    a. To provide a comfortable work environment for employees.

    b. To show when a surface is soiled and when it has been properly cleaned.

    c. To decrease accidents and waste due to worker error.

    d. To reduce glare that causes eye fatigue.

10. Which of the following statements about ventilation is **false**?

    a.   Ventilation is typically provided by means of a mechanical exhaust hood system.

    b.   Ventilation hood systems must be designed and constructed to prevent grease or condensation from dripping onto food and food contact surfaces.

    c.   The hood should be equipped with filters or other grease catching devices to prevent drippage into food.

    d.   Intake and exhaust air ducts do not need cleaning if filters or other grease catching devices are provided to collect grease and other condensation.

Answers to the multiple choice questions are provided in Appendix A.

References/Suggested Readings

Cichy, R. F. (1994). *Quality Sanitation Management.* Educational Institute of the American Hotel and Motel Association. East Lansing, MI.

Food and Drug Administration and Conference for Food Protection (1997). *Food Establishment Plan Review Guide.* Washington, DC.

Khan, M. A. (1991). *Foodservice Operations and Management.* Van Nostrand Reinhold. New York, NY.

Kotschevar, Lendal H. and Margaret E. Terrell (1985). *Foodservice Planning: Layout and Equipment.* Wiley. New York, NY.

Longrèe, K., and G. Armbruster (1996). *Quantity Food Sanitation.* Wiley Interscience. New York, NY.

Knight, J. B., and L. H. Kotschevar (1989). *Quantity Food Production, Planning, and Management.* Van Nostrand Reinhold. New York, NY.

NSF International, Ann Arbor, Michigan.

STANDARD Number 2 - Food Equipment

STANDARD Number 3 - Commercial Spray-Type Dishwashing Machine

STANDARD Number 4 - Commercial Cooking, Rethermalization, and Powered Hot Food Holding and Transport Equipment

STANDARD Number 5 - Water Heaters, Hot Water Supply Boilers, and Heat Recovery Equipment

STANDARD Number 6 - Dispensing Freezers

STANDARD Number 7 - Food Service Refrigerators and Storage Freezers

STANDARD Number 8 - Commercial Powered Food Preparation Equipment

STANDARD Number 12 - Automatic Ice Making Equipment

STANDARD Number 13 - Refuse Compactors

STANDARD Number 18 - Manual Food and Beverage Dispensing Equipment

STANDARD Number 20 - Commercial Bulk Milk Dispensing Equipment

STANDARD Number 21 - Thermoplastic Refuse Containers

STANDARD Number 26 - Pot, Pan, and Utensil Washers

STANDARD Number 29 - Detergent and Chemical Feeders for Commercial Spray-Type Dishwashing Machines

STANDARD Number 35 - Laminated Plastics for Surfacing Food Service Equipment-

STANDARD Number 36 - Dinnerware

STANDARD Number 37 - Air Curtains for Entranceways in Food and Food Service Establishments

STANDARD Number 51 - Plastic Materials and Components Used in Food Equipment

STANDARD Number 52 - Supplemental Flooring

STANDARD Number 59 - Food Carts

CRITERIA C-2 — Special Equipment and/or Devices

Pannell, Dorothy Van Egmond (1990). *School Foodservice Management*. Van Nostrand Reinhold. New York, NY.

Scriven, C., and J. Stevens (1989). *Manual of Equipment and Design for the Foodservice Industry*. Van Nostrand Reinhold. New York, NY.

Scriven, C., and J. Stevens (1989). *Food Equipment Facts: A Handbook for the Foodservice Industry*. Van Nostrand Reinhold. New York, NY.

West, B. B., et al. (1986) *Food Service in Institutions*. Macmillan Publishers. New York, NY.

Selected Web Sites

NSF International
www.nsf.org

Underwriters Laboratories Inc.
www.ul.com

Central Restaurant Products
www.centralrestaurant.com

Champion Industries
www.championindustries.com

Duke Manufacturing Company
www.dukemfg.com

Bowerman Associates
www.ebowerman.com

Hobart Corporation
www.hobartcorp.com

Katchall Industries International
www.KatchAll.com

Shat-R-Shield Inc.
www.shat-r-shield.com

National Restaurant Association
www.restaurant.org

The Food Marketing Institute
www.fmi.org

Gateway to Government Food Safety Information
www.foodsafety.gov

United States Department of Agriculture (USDA)
www.usda.gov

United States Food and Drug Administration (FDA)
www.fda.gov

USDA Foods Safety and Inspection Service (FSIS)
www.fsis.usda.gov

7 CLEANING AND SANITIZING OPERATIONS

Hepatitis A Outbreak Takes Toll on Bar Patrons

Seventeen people contracted Hepatitis A from food they consumed at a local bar. Of the seventeen cases, at least two were hospitalized and none died.

Investigators from the local health department discovered a number of problems with the food handling techniques and personal hygiene practice used by employees at the bar. Of particular concern was a lack of handwashing prior to food preparation; the transfer of salad materials by bare hands from a bulk container onto serving plates rather than using proper utensils such as tongs; rinsing hands in the third compartment of the 3-compartment sink rather than washing at handwashing stations where soap was available; employees frequently did not wash their hands when they returned from break; plates were stacked when transporting them from the cook line to customers; waitstaff bussed soiled glasses with their fingers inside the glasses and did not wash their hands afterwards. Interviews with the staff at the bar revealed that at least 11 employees have worked recently while ill with diarrhea and/or vomiting

Learning Objectives

After reading this chapter you should be able to:

▲ Recognize the difference between cleaning and sanitizing

▲ Identify the different processes that can be used to clean and sanitize equipment and utensils in a food establishment

▲ Identify the primary steps involved in manually and mechanically cleaning and sanitizing equipment and utensils

▲ Describe the factors that affect cleaning efficiency

▲ Identify the procedures used to clean environmental areas in a food establishment.

Essential Terms

Cleaning agent	Sanitizer
Clean-in-place (CIP)	Chlorine
Detergent	Iodophores
In-place sanitizers	Quaternary Ammonium Compounds
Sanitize	Selectivity
	Soap

Principles of Cleaning and Sanitizing

Cleaning and sanitizing are important activities in a food safety program. Food contact surfaces of equipment and utensils must be cleaned and sanitized at regular intervals to prevent contamination and cross contamination. Equipment and utensils used to prepare potentially hazardous foods are food-contact surfaces which must be washed and sanitized:

▲ When switching from raw to cooked or ready-to-eat foods

▲ After any interruption of an operation that could cause contamination.

Cleaning and sanitizing are two distinct processes used for very different purposes. **Cleaning is the *physical removal of soil and food residues* from surfaces of equipment and utensils. Sanitizing** (sometimes called sanitization) **is the treatment of a surface that has been previously cleaned to reduce the number of disease-causing microorganisms to safe levels** (Figure 7.1). The equipment and supplies used for cleaning are different from those used for sanitizing.

Effective cleaning consists of four separate events:

1. A detergent or other type of cleaner is brought into contact with the soil

2. The soil is loosened from the surface being cleaned

3. The loosened soil is dispersed in the wash water

4. The dispersed soil is rinsed away along with the detergent to prevent it from being redeposited onto the clean surface.

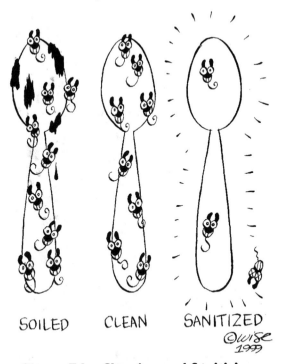

SOILED   CLEAN   SANITIZED
©wise
1999

Figure 7.1—Cleaning and Sanitizing

Removal of Food Particles

Scrape and flush large food particles from equipment and utensils before the items are placed in a cleaning solution (Figure 7.2). Spray the equipment and utensils with warm water. Avoid using very hot water or steam because they tend to "bake" food particles on the surface of equipment and utensils and that makes cleaning more difficult.

Application of Cleaning Agents

A cleaning agent is a chemical compound formulated to remove soil and dirt. There are many methods of applying cleaning agents and solutions to the surfaces of equipment. Cleaning agents typically include an acid or alkaline detergent and may include degreasers, abrasive materials or a sanitizer. Effectiveness and the economy of the method generally dictate its use.

Figure 7.2—Pre-Flushing Makes Washing Easier

Soaking

Small equipment, equipment parts, and utensils may be immersed in cleaning solutions in a sink. By soaking equipment or utensils for a few minutes before scrubbing, you will increase the effectiveness of manual and mechanical dishwashing.

Spray Methods

Spray cleaning solutions on equipment surfaces can use either fixed or portable spray units that use hot water or steam. These methods are extensively used in meat departments of grocery stores and in food processing plants.

Clean-in-Place Systems

The **clean-in-place** method is an automated cleaning system generally used in conjunction with permanent-welded pipeline systems. The strength and velocity of the cleaning solution moving through the pipes are chiefly responsible for soil removal in clean-in-place (CIP) operations.

Abrasive Cleaning

Abrasive type powders and pastes are used to remove soil that is firmly attached to a surface. Always rinse these cleaners completely and avoid scratching the surface of equipment and utensils. Abrasive type cleaners are not recommended for use on stainless steel surfaces. Never use metal or abrasive scouring pads on food-contact surfaces because small metal pieces from the pads may promote corrosion or may be picked up in food to become a physical hazard.

Rinsing

Immediately after cleaning, thoroughly rinse all equipment surfaces with hot, potable water to remove the cleaning solution. This very important rinse step is necessary because the product or detergent used for washing can interfere with the germ-killing power of the sanitizer.

Factors Affecting Cleaning Efficiency

Many factors affect the efficiency of a cleaning process in terms of how well and how easily soil is removed. Among the most important factors are:

▲ Type of soil to be removed

▲ Water quality

▲ The detergent or cleaner to be used

▲ Water temperature

▲ Water velocity or force

▲ Time detergent remains in contact with the surface

▲ The concentration of cleaner.

Each of these factors is important for effective cleaning and can be modified to meet the needs of a particular type of problem or set of conditions.

Type of Soil to Be Removed

Soil may consist of:

▲ Food deposits (proteins and carbohydrates)

▲ Mineral deposits (salts)

▲ Microorganisms (bacteria, viruses, yeasts, and molds)

▲ Fats and oils

▲ Dirt and debris.

When you know the type and condition of the soil you are working with, it is easier to determine the physical and chemical methods best suited to remove that soil effectively and economically.

Water Quality

Water is the primary component of cleaning materials used in food establishments. **The water supply serving food establishments must be safe to drink (potable). Water must be free from harmful microorganisms, chemicals, and other substances that can cause disease.** While the water supply must be safe to drink, it may contain substances that affect hardness, taste, and odors. Therefore, adjust cleaning agents to fit the characteristics of your water supply and type of operation.

"Hard" water is caused by dissolved salts of calcium, magnesium, and iron. Water hardness reduces the effectiveness of detergents and leaves "lime" scale deposits on the surface of equipment. The degree of water hardness varies considerably from place to place. For effective cleaning, hard water requires softening. If the water is not softened, the cleaner must contain a sequestering (softening) agent that inactivates the iron and manganese without settling them out.

If there is an interruption in the water supply to a food establishment, the facility should cease operations or serve only pre-prepared food using single-use disposable utensils.

Detergents and Cleaners to Be Used

A **detergent** is defined as a cleaning or purifying agent—a solvent. The origin of the word is from the Latin, "detergeo," meaning "to wipe away." Water acts as a detergent when soils are readily soluble. However, we can improve the cleansing action of water by adding soap, alkaline detergents, acid detergents, degreasers, abrasive cleaners, detergent sanitizers, or other cleaning agents to it.

Soaps

Bar **soaps** are cleaning agents made by chemical reaction of alkali on fats or fatty acids. Soaps are only used as cleaners in food establishments in a limited way. In soft water, they are effective for washing hands. However, most soaps do not dissolve in cold water and are not compatible with some sanitizers. In hard water, soaps form troublesome precipitates and films that lose their cleaning power and break down into fatty acids. Bar soaps have largely been replaced by synthetic detergents in today's food establishments.

Alkaline Detergents

An alkali is the principal detergent ingredient of most cleaners. Alkalis combine with fats to form soaps, and with proteins to form soluble compounds easily removed by water. Some alkalis are good buffers which enhance detergency. However, in varying degrees, alkalis corrode aluminum, galvanized metal, and tin.

Sodium hydroxide (caustic soda or lye) is the strongest and cheapest of strong alkalis. It has high detergency but is also highly corrosive for nearly all surfaces, including skin. It rinses with difficulty. Wetting agents improve rinsing and wetting while inhibiting corrosion.

Mild alkalis have moderate dissolving power and are much less corrosive than strong alkalis. Some skin contact with mild alkalis is possible. Sodium carbonate competes with sodium hydroxide as the cheapest of the alkalis. Sodium carbonate is safer and less corrosive and therefore is a common ingredient in cleaners.

Alkaline detergents are good general-purpose cleaners that are commonly used to clean equipment, utensils, floors, walls, and ceilings in food establishments.

Acid Detergents

Acid cleaners dissolve mineral deposits, such as calcium and magnesium precipitates, from equipment surfaces. Acid detergents are frequently used to remove food and hard water deposits from the surfaces of equipment and utensils.

The acids may be categorized into two groups: inorganic and organic. The inorganic acids, also called mineral or strong acids, include hydrochloric (HCl), sulfuric (H_2SO_4), nitric (HNO_3) and phosphoric (H_3PO_4). The strong acids are extremely corrosive to metals and are used only in special cases. Cleaners to remove milk films often contain nitric acid, and those to remove hard water films often contain phosphoric acid. The organic acids are the active ingredients of a wide variety of acid cleaners. The organic acids are not as corrosive to metals and are less irritating to the skin.

Degreasers

Degreasers are specialty products that remove grease and greasy or oily soil. Surfactants, their basic ingredients, penetrate and break up grease and oil. Degreasers are designed more for hard surfaces than for fabrics. They may be used for pre-treatment or as the sole cleaning agent, but their use should always be followed by rinsing.

Abrasives

Abrasives, when mixed with a detergent, are useful for jobs that require scrubbing, scouring, or polishing. Such naturally occurring mineral abrasives as pumice, quartz, and sand are ground into small particle size and supply scouring and polishing action to cleansers, hand soaps, and soap pads.

Abrasives should be used with care since they can cause scratches on metal surfaces, including stainless steel. In addition, the particles of the cleaner, along with metal particles scraped from equipment, may also contaminate food.

Detergent Sanitizers

As the name would suggest, detergent sanitizers are compounds that contain both a detergent and a sanitizer. These products effectively clean and sanitize a food-contact surface, provided a two-step process is used. That is, the detergent sanitizer must be applied to a food-contact surface two times. The first time to clean the surface and a second time to sanitize it. Food-contact surfaces should be thoroughly rinsed to remove chemical residues that may remain on the surface after the cleaning and sanitizing process.

Water Temperature

Heat-stable detergents work best when the water temperature of the solution is between 130°F (54°C) and 160°F (71°C). The temperature of the wash water needs to be hot enough to effectively remove soil, but it should not be so hot that the soil will be "baked" onto the food contact surface. Increased wash water temperatures help decrease the strength of the bonds that hold soil to the surface being cleaned.

Some detergents are designed specifically for cold water. Check the manufacturer's directions to determine the appropriate temperatures. A good rule of thumb is to apply cold-water detergents at tap-water temperature.

Velocity or Force

In manual cleaning procedures, force is applied by "elbow grease." For mechanical and clean-in-place systems, the velocity and force of the cleaning solution helps remove soil and film from food contact surfaces. The importance of velocity and force on cleaning decreases as the effectiveness of the detergent increases. In other words, less "elbow grease" will be required when the detergent is formulated to effectively remove soil from the surfaces of equipment and utensils.

Amount of Time the Detergent or Cleaner Remains in Contact with the Surface

Cleaning efficiency is increased by use of longer contact times. The amount of scrubbing necessary to remove soil can be decreased by "soaking" items in detergents or cleaners before cleaning.

Concentration of the Detergent or Cleaner

Increasing the strength of a detergent increases the reaction rate and magnifies its cleaning power. However, there is an upper limit beyond which higher concentrations neither increase nor decrease efficiency. Always follow the manufacturer's recommendations for detergent concentration. Using too much detergent is a waste of money.

Sanitizing Principles

Heat and chemicals are the two types of sanitizers most commonly used in food establishments. **Sanitizers destroy disease-causing organisms which may be present on equipment and utensils even after cleaning.** Sanitization is not sterilization, because some bacterial spores and a few highly resistant vegetative cells generally survive.

In all instances, a food contact surface must be cleaned and then thoroughly rinsed to remove loosened soil and detergent residues that tend to inhibit the sanitizer's action.

Heat Sanitizing

Heat has several advantages over chemical sanitizing agents because it:

▲ Can penetrate small cracks and crevices

▲ Is non-corrosive to metal surfaces

▲ Is non-selective to microbial groups

▲ Leaves no residue

▲ Is easily measurable.

Heat destroys vegetative bacteria cells by disrupting some of the protein molecules in the cells. Moist heat is much more efficient in killing microorganisms than dry heat. Heat sanitization is used in both manual and mechanical warewashing operations.

Heat sanitizing in manual warewashing operations involves immersing cleaned equipment and utensils for at least 30 seconds in hot water that is maintained at 171°F (77°C) or above. A properly calibrated thermometer must be kept on hand to routinely check the temperature of the sanitizing water. Use dish baskets or racks to lower equipment and utensils into the sanitizing water. This allows employees to cover the items completely with the hot water without being scalded. Use extreme care with this process. Enforce procedures that protect the safety of the employees involved in manual dishwashing operations.

Sanitizing with hot water can also be performed in mechanical warewashing equipment. The *Food Code* requires a temperature of not less that 180°F (82°C) for the hot water sanitizing rinse in mechanical warewashers. However, in single tank, stationary rack, and single-temperature machines, the final rinse temperature must be at least 165°F (74°C). The temperature of the hot water sanitizing rinse in a mechanical warewashing machine must not be more than 194°F (90°C) at any time.

Steam is an excellent agent for treating pieces of food equipment. Steam flow in cabinets should be sufficient to achieve 171°F (77°C) for at least 15 minutes or 200°F (94°C) for at least 5 minutes.

When using heat for sanitization, it is the temperature at the utensil surface that is most important to insure proper destruction of microorganisms. The temperature at the utensil surface can be measured by using irreversible heat-sensitive labels or tapes that are attached directly to equipment and utensils by a self-adhesive. The silver labels will turn black when the required sanitization temperature is reached. An example of a T-stick is presented in Figure 7.3.

Figure 7.3—A T-Stick Can Be Used to Measure the Sanitizing Temperature in Dishwashing Machines

Another popular device for measuring the temperature of hot water sanitizers is the "maximum registering" or "holding" thermometer. This type of thermometer will continue to hold the highest temperature measured until shaken down like a medical thermometer. An example of a maximum registering thermometer is presented in Figure 7.4

Chemical Sanitizing

To chemically sanitize, one either immerses an object in a sanitizing solution or by swabbing, brushing, or pressure spraying a sanitizing solution directly on the surface to be sanitized.

The effectiveness of a chemical sanitizer weakens as bacteria and other microorganisms are destroyed. Also, water from the wash and rinse stages of the warewashing process will dilute the sanitizer. Keep a chemical test kit or test strips on hand to permit personnel to routinely check the strength of the sanitizing solution. Replace the sanitizing solution in the sink whenever it is contaminated or if the concentration falls below the minimum recommended by the chemical manufacturer.

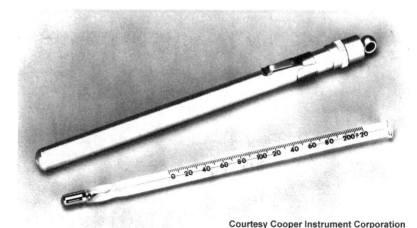

Figure 7.4—Maximum Registering Thermometer

There are a wide variety of chemicals whose properties destroy or inhibit the growth of microorganisms. However, many of these chemicals are not suitable for use on food-contact surfaces because they may corrode, stain, or leave a film on the surface. Others may be toxic or too expensive for practical use. Therefore, the chemical sanitizing agents discussed in this section will be limited to those agents commonly used in food establishments. If a chemical sanitizer other than chlorine, iodine, or a quaternary ammonium compound is used, it must be applied in accordance with the manufacturer's use direction included in the labeling.

Chlorine

Chlorine is a chemical component of hypochlorites which are commonly used as chemical sanitizers in food establishments. Hypochlorites offer many advantages as sanitizers for food establishments. The chief advantages are that they:

▲ Control a wide range of microorganisms through germicidal action

▲ Deodorize *and* sanitize

▲ Are non-toxic to humans when used at recommended concentrations

▲ Are colorless and non-staining

▲ Are easy to handle

▲ Are economical to use.

Hypochlorites are available as powders or liquids. Calcium hypochlorite is generally available in powder form and contains 70% available chlorine. Sodium hypochlorite, more commonly called household bleach, comes in a liquid form and contains between 1 and 15% available chlorine. Hypochlorites release hypochlorous acid in solution. It is the hypochlorous acid that provides chlorine's germ-killing power.

The germicidal effectiveness of chlorine-based sanitizers depends, in part, on water temperature and pH of the solution. According to the *Food Code*, a chlorine solution must satisfy the minimum concentration requirements presented in Figure 7.5.

Maximum Concentration	Minimum Temperature	
ppm	pH 10 or less	pH 8 or less
25	120°F (49°C)	120°F (49°C)
50	100°F (38°C)	75°F (24°C)
100	55°F (13°C)	55°F (13°C)

Figure 7.5—Minimum Concentration of Chlorine-Based Sanitizers

The *Food Code* recommends that food contact surfaces be exposed to a 50-ppm chlorine sanitizer for at least seven seconds. To qualify for the seven-second contact time, the sanitizer solution must have a pH of 10 or less and a water temperature of 100°F (38°C) **or** a pH of 8 or less and a temperature of at least 75°F (24°C). An exposure time of 10 seconds is required if the strength, pH, and temperature requirements described above are not met. Always leave equipment and utensils in contact with the sanitizer long enough to significantly reduce the number of germs that remain on the surface after cleaning. State and local food codes may require that equipment and utensils be immersed in chlorine sanitizers for a longer period of time. Consult with the food regulatory personnel in your area to determine the sanitizing requirements for your establishment.

The effectiveness of hypochlorites is reduced by organic matter (i.e., food soil) even in small amounts. Therefore, rinse the detergent off equipment and utensils before immersing them in the sanitizing solution. A chlorine test kit or strips must be available to measure the concentration of the chlorine at least once per hour. Replace the sanitizer whenever the concentration falls below the required strength.

Iodine

Iodine is chemically related to chlorine and has long been used to kill germs. The iodine-containing sanitizers commonly used in food establishments are called **iodophors**. Iodophors are effective against a wide range of bacteria, small viruses, fungi, and the spores of several disease-causing bacteria. Iodophors kill more quickly than either chlorine or the quaternary ammonium compounds.

Iodophors function best in acidic solutions at temperatures between 75° and 120°F (24° and 49°C). It is recommended that iodophors be applied at 12.5 parts per million in immersion sanitizing and 25 parts per million in swab and spray applications. Expose a surface to an iodophor for at least 30 seconds to ensure proper sanitizing.

Iodophors are less influenced by organic matter than are hypochlorites and quaternary ammonium compounds. However, they are more expensive to use, and they will discolor and stain some surfaces such as silver, silverplate, and copper. Iodophors are also slippery and harder to handle than hypochlorites.

Quaternary Ammonium Compounds (quats)

Quaternary ammonium compounds (quats) are ammonia salts that are used as chemical sanitizers in food establishments.

Quats are effective sanitizers, but are selectively used for certain types of bacteria. The *Food Code* recommends using quats at 200 parts per million (ppm) for immersion sanitizing of food contact surfaces. Quats are non-corrosive and non-irritating to skin and have no taste or odor when used in the proper dilution. Quats work well at all temperatures above 75°F (24°C). They are effective in a wide pH range (although they are most effective in slightly alkaline water). The recommended contact time for quats is at least 30 seconds.

Quats should be used only in water with 500 ppm hardness or less, or in water having a hardness no greater than specified by the manufacturer's label.

At concentrations above 200 ppm, quats will leave a residue on the surface of an item. This is undesirable for food-contact surfaces. It is generally recommended that quats not be used at concentrations above 200 ppm. Follow the manufacturer's directions for proper uses and concentration of quats. Also consult with the regulatory agency in your jurisdiction about the uses of these and other chemical sanitizers.

Factors that Affect the Action of Chemical Sanitizers

The effectiveness of chemical sanitizers is affected by many different factors. Following are some of the most important factors to consider.

▲ **Contact of sanitizer**—in order for a chemical to react with microorganisms, it must achieve intimate contact.

▲ **Selectivity of sanitizer**—certain sanitizers are non-selective in their ability to destroy microorganisms whereas others exhibit a degree of selectivity. Chlorine is relatively non-selective. However, both iodophors and quaternary compounds have a selectivity which may limit their application.

▲ **Concentration of sanitizer**—in general, increasing the concentration of a chemical sanitizer proportionately increases the rate of microbial destruction. But there are certain limitations. The increased activity only extends to a certain maximum, and then levels off so that any further increase in concentration has no advantage. In rare instances, the activity actually decreases with increased concentration. Concentration limits have been set for each type of sanitizer. *More is not always better, and high concentrations of sanitizers can be toxic.* Always follow the manufacturer's label use instructions to ensure peak effectiveness of chemical sanitizers.

▲ **Temperature of solution**—all of the common sanitizers increase in activity as the solution temperature increases. This is partly based on the principle that chemical reactions in general are speeded up by raising the temperature. The recommended range of water temperatures for chemical sanitizing solutions is between 75°F (24°C) and 120°F (49°C). Water temperatures above 120°F (49°C) should be avoided when using chlorine and iodine. At high temperatures, the potency of these sanitizers is lost by its evaporation into the atmosphere.

▲ **pH of solution**—water hardness can effect the pH of water which exerts a significant influence on most sanitizers.

Quaternary compounds react differently to changing pH, depending on the type of organism being destroyed. Chlorine and iodophors generally decrease in effectiveness with an increase in pH. Most soaps and detergents are alkaline with a pH between 10 and 12. That is why the soap or detergent must be rinsed off before a surface is sanitized.

▲ **Time of exposure**—allow sufficient time for chemical reactions to destroy the microorganism. The amount of exposure time depends on the preceding factors as well as the size of the microbial populations and their susceptibility to the sanitizer.

Mechanical Dishwashing

The mechanical dishwashing machine has been used in food establishments for many years to clean and sanitize multiple-use equipment and utensils. A dishwashing machine is designed to clean and sanitize large quantities of equipment and utensils that will fit into the machine and that have no electrical parts. When properly operated and maintained, machine dishwashing is more reliable than manual dishwashing for removing soil and bacteria from equipment and utensils.

Mechanical Dishwashing Process

The mechanical dishwashing process employs eight separate steps to wash and sanitize equipment and utensils (Figure 7.6). A summary of the mechanical process follows. You must:

1. Pre-scrape and pre-flush soiled equipment and presoak utensils to remove visible soil.

2. Rack equipment and utensils so that wash and rinse waters will spray evenly on all surfaces and the equipment will freely drain.

3. Wash equipment and utensils in a detergent solution that satisfies the temperature requirements prescribed in Figure 7.7.

4. Rinse equipment and utensils in clean water at a temperature consistent with the type of dishwashing machine being used (Figure 7.7).

5. Rinse equipment and utensils in a fresh hot water sanitizing rinse between 180° and 194°F (82° and 90°C), except for a single tank, stationary rack, or single temperature machine,

where the final rinse may not be less than 165°F (74°C). The recommended final rinse temperature for low-temperature chemical sanitizing dishwashing machines is 120°F (49°C) or less.

6. Air-dry equipment and utensils.

7. Store clean and sanitary items in a clean, dry area where they are protected from contamination.

8. Clean and maintain the machine to keep it in proper working condition.

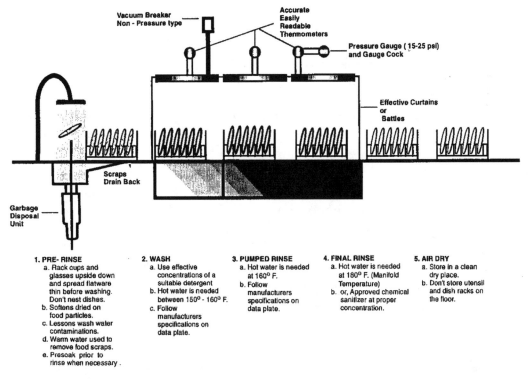

Figure 7.6—Diagram of Machine Dishwashing System

The majority of commercial spray-type dishwashing machines on the market today will work efficiently. The major problems with this type of equipment are operational and require ongoing supervision and surveillance. Selection of a particular machine for a given food establishment requires knowledge of the demands to be placed on the machine. This includes the type of utensils to be washed and the quantity of equipment and utensils to be processed during peak periods.

Provide sufficient counter or table space for the accumulation of soiled equipment and utensils and keep them separate from items that are clean and sanitary. Scrape food debris into a garbage disposal or garbage can. Freshly deposited soil is the material that remains on equipment and utensils after meal preparation or a meal itself. This is the easiest type of soil to remove because washing usually occurs while the soil is still moist.

Dried deposits of soil result when dirty dishes and utensils stand for a long period of time or are exposed to high temperatures. This allows fresh soft soil to dry, harden, and form a crusty deposit that is hard to remove. Dried soil frequently requires pre-soaking to soften and loosen it from the surface.

Once items are pre-flushed and pre-scraped, position them in a rack or on a conveyor so that wash and rinse water and the sanitizer will spray evenly on all surfaces. Glasses and small equipment are racked with their bottom side up. Dishes are racked on their edge and eating utensils are placed in baskets with the mouth piece up. After the utensils are washed, rinsed, and sanitized, invert them into another basket with the handles up and repeat the warewashing process.

The temperature of the wash water (and power rinse if applicable) is extremely important in providing effective sanitizing of equipment and utensils. Minimum wash and rinse water temperatures are provided in Figure 7.7. Equipment and utensils should be exposed to the wash water for at least 40 seconds.

Some machines provide a power rinse which removes soil and detergent residues. The final rinse is designed to sanitize at a temperature of not less than 165°F (74°C) for at least 30 seconds in single tank, stationary rack, single temperature machines. Sanitizing in all other machines requires a final rinse temperature between 180° and 194°F (82° and 90°C) for at least 30 seconds. The flow pressure of the hot water sanitizing rinse must be between 15 and 25 pounds per square inch (psi) to ensure the water makes good contact with the surface of the items being sanitized.

Wash and power rinse water temperatures less than those prescribed in the Food Code will not effectively sanitize equipment and utensils even when the final rinse (sanitizing) temperature is properly maintained. This is because effective sanitizing of dishes is the result of the cumulative temperature effects of wash, power rinse, and final rinse waters. Dish machine operators must be careful not to contaminate cleaned and sanitized equipment and utensils by touching them with soiled hands.

Type of Machine	Wash Temperature	Rinse Temperature
Single tank, stationary rack, single temperature machine	165°F 74°C	165°F 74°C
Single tank, conveyor, dual temperature machine	160°F 71°C	180°F 82°C
Single tank, stationary rack, dual temperature machine	150°F 66°C	180°F 82°C
Multitank, conveyor, multi-temperature machine	150°F 66°C	160°-180°F 71°- 82°C
Chemical sanitizing machine	120°F 49°C	120°F 49°C

Figure 7.7—Minimum Wash and Rinse Temperatures of Mechanical Dishwashing Machines
(Source: *Food Code*)

Cleaned and sanitized equipment and utensils *must be air dried* before storage (Figure 7.8). Soiled, multi-use cloth towels can recontaminate cleaned and sanitized equipment and utensils. The *Food Code* prohibits cloth-drying of equipment and utensils, with the exception that air-dried utensils can be "polished" with cloths that are clean and dry.

Cleaned and sanitized equipment and utensils must be stored in such a way as to protect them from contamination (Figure 7.9). Cabinets and carts used to store cleaned and sanitized equipment and utensils may not be located in locker rooms and toilet rooms and under sewer lines that are not shielded to catch drips, leaking water lines, open stairwells, or under other sources of contamination.

Figure 7.8—Air-Dry Utensils

Figure 7.9—Clean Dish Storage

Manual Dishwashing

Manual warewashing is typically performed in a sink that has at least three compartments (Figure 7.10). The first compartment is used for washing, the second for rinsing, and the third compartment is used for sanitizing. Each compartment must be large enough to accommodate immersion of the largest equipment and utensils used in your food establishment. If equipment or utensils are too large for the warewashing sink, they must be cleaned and sanitized using a mechanical dishwashing machine or alternative manual warewashing equipment, such as high-pressure detergent sprayers, low- or line-pressure spray detergent foamers, or other task-specific cleaning equipment.

Wash-Rinse-Sanitize

Figure 7.10—Manual Dishwashing

The manual warewashing sink must be equipped with sloped drain boards or dish tables of adequate size to store soiled items prior to washing, and clean items after they have been sanitized. Hot and cold potable water must be supplied to each compartment of the sink, and the sink should be cleaned and sanitized prior to use.

Equipment and utensils should be scraped, flushed, and, when necessary, presoaked to remove food residues and other soil. Proper pre-flushing and pre-scraping helps preserve the cleanliness of the wash water in the first compartment of the sink.

Equipment and utensils are washed in the first compartment with warm water and an effective cleaning agent. Washing removes visible food particles and grease. The correct amount of detergent, based on the quantity of water used in the compartment, should always be used. The temperature of the wash solution must be maintained at not less than 110°F (43°C) unless a different temperature is specified by the manufacturer of the cleaning agent.

A cloth, nylon brush, or other approved cleaning device should be used to loosen and remove soil.

Equipment and utensils are rinsed in clean, warm water in the second compartment. Rinsing removes cleaning agents, soap film, remaining food particles, and abrasives that may interfere with the sanitizer agent. For best results, the rinse water must be kept clean and the temperature should be at least 120°F (49°C).

Items are sanitized in the third compartment by submerging them in hot water or a chemical sanitizer solution. **When hot water is used as a sanitizer, the temperature of the water must be maintained at 171°F (77°C) or above at all times.** Items must be immersed in the hot water for at least 30 seconds. Chemical sanitizers are frequently preferred in food establishments because they do not require large amounts of hot water, which saves energy. The *Food Code* recommends that items being chemically sanitized be submerged in the sanitizer solution for at least 10 seconds for a chlorine compound and 30 seconds for other chemical sanitizers. Most state and local food codes require immersion in all chemical sanitizers for at least 60 seconds. You should consult the regulatory authority in your jurisdiction to find out what rules apply in your area. **The recommended range of water temperatures for most chemical sanitizing solutions is between 75°F (24°C) and 120°F (49°C). Except for quats, water temperatures above 120°F (49°C) must be avoided because at high temperatures the potency of the sanitizer is lost by evaporation into the air.** Soil and other deposits can shield bacteria from chemical sanitizing solutions. Therefore, always make certain that all surfaces are properly cleaned before they are sanitized.

As with mechanical warewashing, items that have been manually cleaned and sanitized must be air dried before they are put into storage. Avoid wiping cleaned and sanitized equipment and utensils with cloths or towels because this can recontaminate the sanitized surfaces.

Cleaned and sanitized equipment and utensils must be stored in such a way as to protect them from contamination. Cabinets and carts used to store cleaned and sanitized equipment and utensils may not be located in locker rooms and toilet rooms and under sewer lines that are not shielded to catch drips, leaking water lines, open stairwells, or under other sources of contamination.

The effectiveness of the manual warewashing process depends to a great degree on the condition of the wash, rinse, and sanitize water. When the cleaning agent disappears in the first compartment, the wash solution should be completely drained and replaced with fresh detergent. When detergent builds up in the second compartment, the fresh water rinse should be drained and replaced with clean, fresh water. When the sanitizer in the third compartment is depleted, it should be drained completely and a new solution made. The concentration of a chemical sanitizer solution must be tested periodically with a chemical test kit or strips to make sure that it remains at the strength required by the regulatory agency in your jurisdiction. Obtain test strips from the manufacturer of your sanitizer (see Figure 7.11). These strips provide a color comparison to indicate the strength of the sanitizer.

Figure 7.11—Measuring the Concentration of Sanitizers Using Test Strips

Cleaning Fixed Equipment

Some equipment such as soft-serve yogurt machines, mixers, slicers, and grinders cannot be cleaned and sanitized using either mechanical or conventional manual warewashing processes (Figure 7.12). This equipment should be disassembled to expose food contact surfaces

Courtesy of Hobart Corp.

Figure 7.12—Meat Grinder

to cleaning and sanitizing agents The basic steps for cleaning fixed equipment are:

1. To avoid electrocution and other injury, always disconnect power to the equipment before disassembling it for cleaning.

2. Equipment must be disassembled as necessary to allow access of the detergent solution to all parts.

3. Use a plastic scraper to clean equipment parts and remove food debris that has accumulated under and around the equipment. Scrape all debris into the trash.

4. Carry the parts that have been removed from the equipment to the manual warewashing sink where they will be washed, rinsed, and sanitized.

5. Wash the remaining parts of the equipment using a clean cloth, brush, or scouring pad and warm, soapy water. Clean from top to bottom.

6. Rinse thoroughly with fresh water and a clean cloth.

7. Swab or spray a chemical sanitizing solution, mixed to the manufacturer's recommendations, onto all food-contact surfaces.

8. Allow all parts to drain and air dry.

9. Reassemble the equipment.

10. Resanitize any food-contact surface that might have been contaminated due to handling when the equipment was being reassembled.

Large equipment such as preparation tables and band saws can be cleaned by using a foam or spray method. In this process, detergents and degreasers, fresh water rinse, and a chemical sanitizer are applied using foam or spray guns. The hoses, feed lines, and nozzles that make up the foam or spray unit should be in good condition and attached properly.

The bucket method is often used to clean equipment that could be damaged by pressure spraying or immersion in the manual warewashing sink. This system employs separate buckets for washing, rinsing, and sanitizing. Three buckets are required when sanitizing with a clean cloth and sanitizer solution. Only two buckets (one for wash and another for rinse) are required if the sanitizer is sprayed on the equipment from a spray bottle.

The clean in place (CIP) method is used for equipment that is designed to be cleaned and/or sanitized by circulating chemical solutions through the equipment. An example of this type of equipment is a soft-serve ice cream or yogurt machine (Figure 7.13). The basic steps in the clean-in-place method are:

1. Empty food product and waste from the equipment.

2. Disconnect the power to the equipment.

3. Disassemble if parts are removable and cleanable in the manual warewashing sink.

4. Clean and sanitize the removable parts using the three-compartment warewashing sink and process.

5. Clean and sanitize the main unit by circulating wash, rinse, and sanitizing solutions through it. Combination detergent sanitizers can also be used on equipment that is designed for CIP cleaning.

6. Reassemble parts that were removed from the main unit.

7. Re-sanitize by circulating a manufacturer approved sanitizing solution through the equipment

Courtesy of SaniServ

Figure 7.13—Soft-Serve Ice Cream Machine

The cleaning frequency for kitchen equipment depends on the location and product processed. In most operations, the equipment should be cleaned at least once every four hours or between different types of products. Always check the manufacturer's instructions for recommended cleaning frequency and procedures.

Many food operations use wiping cloths to clean up spills. Wiping cloths should be stored in an in-place sanitizing solution in order to reduce microbial growth on the wiping cloths between uses (Figure 7.14). The in-place sanitizing solution can contain chlorine, iodine, or quarternary ammonia compounds. Quarternary ammonia compounds are preferred because they are capable of killing germs for a longer period of time. Wiping cloths must be changed regularly, and they must be used for no other reason except for removing spills. Cloths used for wiping floors, raw food preparation areas, and equipment must be kept separate. The *Food Code* requires wiping cloths to be laundered daily.

Many food establishments use a variety of single service items. These are typically paper or plastic dishes, cups, and eating utensils that are used one time and then discarded. Single use items must be stored and dispensed to protect them from contamination from customers and the surrounding environment. Dishes and cups should be stored in boxes or cabinets or upside down and away from overhead plumbing lines to protect them from contamination. Eating utensils should be either wrapped in plastic or stored in an approved dispensing container with the handles up. This will prevent the eating part of the utensil, the one that comes into contact with your customer's mouth, from becoming contaminated.

Figure 7.14—Store and Use Wiping Cloths Properly

Cleaning Environmental Areas

Walls, ceilings, floors, and drains are non-food contact surfaces that must be cleaned regularly to minimize contamination, odors, and insect and rodent infestations. Use only chemicals accepted for food contact equipment when cleaning and sanitizing walls, ceilings, and floors of rooms used for food processing. For these surfaces, it is acceptable to use sanitizers at concentrations higher than those permitted for food-contact surfaces. It is important to prevent dirt and chemicals from falling or splashing onto food-contact equipment or into open containers.

Ceilings

Ceilings should be checked regularly to make certain they are not contaminating food production areas. Ceilings, lights, fans, and covers can be cleaned using either a wet or dry cleaning technique. When wet-cleaning ceilings and fixtures, it is best to use a bucket method to keep water away from lights, fans, and other electrical devices. Whenever possible, disconnect power before cleaning fixtures.

Walls

Some parts of the walls in a food production area, such as those around a sink or food production equipment, should be considered food contact surfaces. These areas should be washed, rinsed, and sanitized whenever the other food contact surfaces in the area are cleaned. Other wall areas are considered non-food contact surfaces and do not need to be cleaned as often.

Walls may be cleaned using either the bucket or spray methods. However, walls should not be spray cleaned if spraying will damage the walls or contaminate exposed equipment and supplies in the area.

Floors

The floor is an important part of production, warewashing, and food storage areas. Floor construction and condition are important factors in maintaining a sanitary operation. Floors in all areas of a food establishment are subject to harsh treatment. Cleaning solutions, hot water, steam, scrubbers, food wastes, and heavy traffic all contribute to floor deterioration.

The cleanliness of floors depends on the ability of an employee to remove soil from the surface of the floor. Floors are usually cleaned using a spray system for washing, rinsing, and sanitizing. If floors will be damaged by spray cleaning, they can be cleaned using the bucket method. All floors should be sloped to drains to help remove soil and the water used to clean floors.

If floor drains are not properly maintained, they can be a source of disease, vermin, and odors. Floor drains under refrigeration cases should be flushed, washed, rinsed, and sanitized.

The steps commonly used to clean these types of floor drains are listed below:

1. Remove the grate or cover over the drain.

2. Clean out waste and other debris from the drain.

3. Use a sprayer or hose to flush the drain and grate or cover.

4. Pour in drain cleaner to break up grease and other waste in the drain.

5. Wash the drain using a brush or water pressure from the sprayer.

6. Rinse the drain with hot water.

7. Pour or spray sanitizer into the drain.

Odor problems from floor drains can be solved by regularly flushing drains with hot water. This should be the last step in the daily cleaning program.

Equipment and Supplies Used for Cleaning

Managers must make certain that employees have the proper equipment and supplies needed to clean food contact and non-food contact surfaces in the food establishment. Some examples of equipment commonly used during cleaning are nylon brushes, cleaning cloths, scouring pads, squeegees, mops, buckets, spray bottles, hoses, and spray or foam guns. Commonly used cleaning supplies include hot water, cleaners, degreasers, and sanitizers. Cleaning equipment and supplies should be stored in areas away from where food and utensils are stored. If possible, cleaning supplies should be stored in a separate room. If space does not permit storage in a separate area, chemicals used for cleaning and pest control must be stored in a locked and labeled cabinet to avoid accidental contamination of food and food contact surfaces.

The Occupational Safety and Health Administration (OSHA) requires that employees have the "Right-to-Know" about the hazardous chemicals to which they may be exposed on the job. Information about hazardous substances in the workplace is provided to employees by way of material safety data sheets (MSDS). Material safety data sheets are developed by chemical manufacturers, and they must be maintained on file and accessible to employees in the food establishment (Figure 7.15).

Material Safety Data Sheet

HEALTH	2*
FLAMMABILITY	0
REACTIVITY	1
Personal Protection	B

I Chemical Identification

NAME:	REGULAR BLEACH	**CAS no.**	N/A
DESCRIPTION:	CLEAR, LIGHT YELLOW LIQUID WITH CHLORINE ODOR	**RTECS no.**	N/A

Other Designations	Manufacturer	Emergency Procedure
EPA Reg. No. 5813-1 Sodium hypochlorite solution Liquid chlorine bleach Liquid Bleach	12 Broadway Broad Oak, CA. 94612	Notify your Supervisor Call your local poison control center or Rocky Mountain Poison Center (303) 573-1014

II Health Hazard Data

*Causes severe but temporary eye injury. May irritate skin. May cause nausea and vomiting if ingested. Exposure to vapor or mist may irritate nose, throat and lungs. The following medical conditions may be aggravated by exposure to high concentrations of vapor or mist: heart conditions or chronic respiratory problems such as asthma, chronic bronchitis or obstructive lung disease. Under normal consumer use conditions the likelihood of any adverse health effects are low. FIRST AID: EYE CONTACT: Immediately flush eyes with plenty of water. If irritation persists, see a doctor. SKIN CONTACT: Remove contaminated clothing. Wash area with water. INGESTION: Drink a glassful of water and call a physician. INHALATION: If breathing problems develop remove to fresh air.

III Hazardous Ingredients

Ingredients	Concentration	Worker Exposure Limit
Sodium hypochlorite CAS# 7681-52-9	5.25%	not established

None of the ingredients in this product are on the IARC, NTP or OSHA carcinogen list. Occasional clinical reports suggest a low potential for sensitization upon exaggerated exposure to sodium hypochlorite if skin damage (eg. irritation) occurs during exposure. Routine clinical tests conducted on intact skin with Liquid Bleach found no sensitization in the test subjects.

IV Special Protection Information

Hygienic Practices: Wear safety glasses. With repeated or prolonged use, wear gloves.

Engineering Controls: Use general ventilation to minimize exposure to vapor or mist.

Work Practices: Avoid eye and skin contact and inhalation of vapor or mist.

V Special Precautions

Keep out of reach of children. Do not get in eyes or on skin. Wash thoroughly with soap and water after handling. Do not mix with other household chemicals such as toilet bowl cleaners, rust removers, vinegar, acid or ammonia containing products. Store in a cool, dry place. Do not reuse empty container; rinse container and put in trash container.

VI Spill or Leak Procedures

Small quantities of less than 5 gallons may be flushed down drain. For larger quantities wipe up with an absorbent material or mop and dispose of in accordance with local, state and federal regulations. Dilute with water to minimize oxidizing effect on spilled surface.

VII Reactivity Data

Stable under normal use and storage conditions. Strong oxidizing agent. Reacts with other household chemicals such as toilet bowl cleaners, rust removers, vinegar, acids or ammonia containing products to produce hazardous gases, such as chlorine and other chlorinated species. Prolonged contact with metal may cause pitting or discoloration.

VIII Fire and Explosion Data

Not flammable or explosive. In a fire, cool containers to prevent rupture and release of sodium chlorate.

IX Physical Data

Boiling point--------------------212°F/100°C (decomposes)
Specific Gravity (H2O=1)------------1.085
Solubility in Water------------------complete
pH----------------------------------11.4

Figure 7.15—Material Safety Data Sheet

Examples of information contained in an MSDS include:

▲ Ingredients

▲ Physical and chemical characteristics

▲ Fire, explosion, reactivity, and health hazard data

▲ How to handle hazardous chemicals safely

▲ How to use personal protective equipment and other devices to reduce risk

▲ Emergency procedures to use if required.

Failure to produce the required MSDS upon demand can result in a substantial fine from OSHA.

Summary

The case presented at the beginning of the chapter illustrates how the Hepatitis A virus can be spread very easily if the basic rules of proper personal hygiene are not followed. Hepatitis A virus is shed in the stool of infected persons. If a food worker does not wash his or her hands after using the toilet, and then handles food that will not be cooked, the food can become a vehicle of infection. In addition, the virus can be spread if employees do not use clean gloves and/or properly cleaned and sanitized equipment and utensils during the production and service of foods. Finally, employees should be excluded from work if actively ill with diarrhea and/or vomiting until their symptoms have gone away.

A good sanitation program starts with a neat, clean, and properly maintained building. A well-maintained facility will help protect food products from contamination, make proper stock rotation easier, prevent entrance of pests, reduce fire hazards, and contribute to the overall safety of the work environment.

The most important step in the development of a food facility's sanitation program is the establishment of sanitation standards and procedures which are designed to correct deficiencies and to assist in emergency situations. Sanitation standards are the facility's policies, and they should reflect the requirements of government regulations.

The effectiveness of a food establishment's sanitation program is directly related to the degree of commitment and concern demonstrated by top management. In both large and small operations, top management must instill the proper attitude among employees and convey incentive and motivation by personal participation and interest in the program.

Proper education and training of food establishment employees are vital to ensure that sanitation requirements are met. Employees responsible for handling food, equipment, and cleanup of the building and grounds must be trained to accept good sanitation as one of their responsibilities. A clean, well-maintained facility can become a source of pride for employees.

For a sanitation program to be effective, cleaning tasks must be scheduled and individual employees assigned to complete the tasks. Employees must be given instructions that include what they are to clean, how to clean it, and what tools and supplies are required to effectively clean it. Cleaning activities must be carried out by properly trained employees, and they must be closely monitored by supervisors to identify and correct problems when they occur.

A good sanitation program is a preventive program which anticipates and eliminates potential hazards before they become serious problems. The value of an effective program is difficult to measure in terms of dollars. Nonetheless, the value of effective sanitation can be evident in the form of diminished problems and increased goodwill with customers and the local regulatory community.

Case Study 7.1

Ted is night manager of the deli at the Regal Supermarket. One evening Ted noticed that there was a large amount of detergent suds in the third compartment of the manual equipment and utensil washing sink. The deli uses chlorine as a sanitizer. When asked about the soap suds in the sanitizing water, the dishwasher indicated this was a common occurrence when there were many things to wash, especially toward the end of the cleaning and sanitizing process.

1. Should Ted be concerned about what he observed? Why?

2. What should Ted tell the dishwasher to do when there is a large amount of detergent in the sanitizer?

3. What could the dishwasher do to prevent detergent from getting into the sanitizer?

Case Study 7.2

On a Sunday morning the food manager of the Shady Rest long-term care facility noticed that the mechanical dishwasher was not producing water hot enough to properly sanitize equipment and utensils that were being run through the machine. It was not possible to get someone to repair the machine until first thing Monday morning.

1. How should the facility wash and sanitize dishes until the dishwasher is fixed?

2. Are there other alternatives that might be used in the interim?

3. Do you think the food service should be suspended at the facility until the machine is repaired? Why or why not?

Case Study 7.3

A Salmonella outbreak involving 107 confirmed and 51 probable cases occurred in Dodge County, Wisconsin. The illness was caused by eating raw ground beef commonly known as steak tartare. Investigators suspect inadequate cleaning of the meat grinder may have been the cause of the problem. Employees of the butcher shop where the ground beef was purchased indicated the parts of the grinder were cleaned and sanitized at the end of each day. However, the auger housing, which was attached with nuts and bolts could not be easily removed for cleaning and sanitizing. Employees indicated they had been instructed not to remove the auger housing for cleaning.

1. What went wrong?

2. How could this outbreak have been prevented?

Answers to case studies are provided in Appendix A.

Discussion Questions (Short Answer)

1. What is the difference between cleaning and sanitizing?

2. Briefly discuss some of the factors that affect cleaning efficiency.

3. What are the main functions of detergents in the cleaning process?

4. What are the basic steps in the manual dishwashing process?

5. Why must cleaned and sanitized equipment and utensils be air dried?

6. How does water temperature affect chemical sanitizers?

7. What are some of the factors that affect the action of chemical sanitizers?

8. What is the best way to measure the strength of a chemical sanitizer?

9. What steps are commonly used to clean fixed equipment that is too large to be placed in a dishwashing machine or three-compartment sink?

10. What purposes do material safety data sheets serve in a food establishment?

Quiz 7.1 (Multiple Choice)

1. What are the proper steps in a manual dishwashing operation after pre-flushing?

   a. Wash, rinse, sanitize, and towel dry.

   b. Rinse, wash, sanitize, and air dry.

   c. Wash, rinse, sanitize, and air dry.

   d. Rinse, wash, sanitize, and towel dry.

2. When sanitizing with hot water in a manual dishwashing operation, the temperature of the water in the final rinse must be maintained at:

   a. 161°F (72°C)

   b. 171°F (77°C)

   c. 181°F (83°C)

   d. 191°F (88°C).

3. Which of the following statements is **false**?

   a. Dishes should be washed in very hot water (above 171°F, 77°C) to effectively remove soil from the surface.

   b. Prescraping helps to remove larger food particles from dishes, which helps keep the wash water clean.

   c. The cleaning compounds used in a food establishment must be tailored to the individual water supply.

   d. Cleaning is a process which removes soil and prevents accumulation of food residues on equipment, utensils, and surfaces.

4. The recommended range of water temperatures for **chlorine** sanitizing solutions is between _____ and _____.

   a. 55° ; 120°F (13° ; 49°C)

   b. 75° ; 120°F (24° ; 49°C)

   c. 95° ; 140°F (35° ; 60°C)

   d. 120° ; 171°F (49° ; 77°C).

5. The strength of the chemical sanitizer in the third compartment of the three-compartment sink must be checked frequently because

   a. If the chemical is too strong, it may ruin fragile dishware.

   b. The chemical strength increases over time which may leave a toxic residue on the surface of equipment and utensils.

   c. The strength of chemical sanitizers may drop off as germs are killed off and the sanitizer is diluted with rinse water.

   d. The strength of the chemical increases as germs are killed off.

6. The first step in developing a cleaning program for environmental areas is to

   a. determine the cleaning needs.

   b. assign cleaning jobs to employees.

   c. obtain the equipment and supplies needed to implement the cleaning.

   d. conduct a training program to teach employees how to clean properly.

7. Which of the following information is **not** provided by a material safety data sheet?

   a. Information about the physical and chemical characteristics of hazardous substances used in the facility.

   b. Information about how to use personal protective equipment and other devices to reduce a worker's risk of injury.

   c. Information about emergency procedures to use when exposed to a hazardous chemical.

   d. Information about approved hazardous waste sites that will accept unused portions of the hazardous materials.

8. Which of the following statements is **false**?

   a. Keeping things clean is the responsibility of every person working in the food industry.

   b. Cleanliness goes beyond the removal of visible soil.

   c. A good sanitation program, if properly organized, will function successfully without the support of top management.

   d. A good sanitation program starts with a neat, clean, and properly maintained building.

Answers to the multiple choice questions are provided in Appendix A.

References/Suggested Readings

Cichy, R.F. (1994). *Quality Sanitation Management*. Educational Institute of the American Hotel and Motel Association. East Lansing, MI.

Educational Foundation of the National Restaurant Association (1998). *ServSafe*. National Restaurant Association, Chicago, IL.

Food and Drug Administration (1999). *1999 Food Code*. United States Public Health Service. Washington, DC.

Gravini, R. B. and D.C. Rishoi (1993). *Food Store Sanitation*. Chain Store Publishing Company. New York, NY.

Selected Web Sites

Bowerman Associates
www.ebowerman.com

Champion Industries
www.championindustries.com

DiverseyLever Inc.
www.diverseylever.com

Duke Manufacturing Company
www.dukemfg.com

Ecolab
www.ecolab.com

Hobart Corporation
www.hobartcorp.com

The Soap and Detergent Association
www.sdahq.org

Steritech
www.stertech.com

NSF International
www.nsf.org

Underwriters Laboratories Inc.
www.ul.com

National Restaurant Association
www.restaurant.org

The Food Marketing Institute
www.fmi.org

Gateway to Government Food Safety Information
www.foodsafety.gov

United States Department of Agriculture (SDA)
www.usda.gov

United States Foods and Drug Administration (FDA)
www.fda.gov

8 ENVIRONMENTAL SANITATION AND MAINTENANCE

High School Cafeteria Reopens after Taking Measures to Correct Rodent and Roach Infestation

Maryland health officials have given conditional approval for a local high school to reopen its lunchrooms, one week after they were shut down because of mice and roach infestation.

Health officials closed two kitchens and a cafeteria at the school after rodent droppings were discovered in the food storage and preparation areas. Breakfast and lunch were prepared at two nearby elementary schools and shipped to the high school during the time the school cafeteria was closed.

In a effort to eliminate the pest infestation problem, school officials took a number of corrective actions. These included sanitizing affected areas, sealing holes in walls and floors, installing screens on windows, and discarding old food.

Health officials reinspected the school cafeteria and said the school could reopen the two lunchrooms provided they installed door sweeps, repaired a sink, and cleaned the storage room. These projects were completed and the cafeteria was allowed to reopen.

Learning Objectives

After reading this chapter, you should be able to:

▲ Describe how the premises of a food establishment can affect customers' opinions about the cleanliness and sanitation of the operation

▲ Identify the criteria that should be used when selecting materials for floors, walls, and ceilings in kitchen, warewashing, and serving areas of a food establishment

▲ Identify the equipment and supplies that must be provided in a properly equipped restroom facility

▲ Identify the main components of a properly equipped handwashing station and explain how conveniently located handwashing facilities contribute to good personal hygiene

▲ Identify the types of plumbing hazards that can have a negative effect on public health and the devices that are commonly employed to prevent these hazards

▲ Explain how proper disposal and storage of garbage and refuse help prevent contamination and pest problems in a food establishment.

Essential Terms

Air gap	Garbage
Backflow	Integrated Pest Management (IPM)
Coving	Refuse
Cross connection	Vacuum breaker

Condition of Premises

The premises of a food establishment are made up of the physical facility, its contents, and the surrounding land or property. The exterior of a food establishment includes the building structure, parking, landscaping, doors and windows. The objective of the exterior design is to attract customers. To accomplish this function, the exterior must make a good impression on the customer (Figure 8.1)

The exterior of the facility must be free of litter and debris that could attract pests and detract from the aesthetics of the facility. Grass and weeds should be regularly mowed to eliminate harborage areas for insects, rodents, and other pests. Walking and driving surfaces should be constructed of concrete, asphalt, gravel, or similar materials to facilitate maintenance and control dust. These surfaces should also be properly graded to prevent rainwater from pooling and standing on parking lots and sidewalks.

Figure 8.1—Keep the Outside of the Premises Clean

Condition of Building

The entrance area to a facility creates a lasting impression. The cleanliness and attractiveness that patrons view upon entering the building influences their overall dining or shopping experience. The entrance should be easy to find and have an inviting appearance. Locate entrance doors for easy access from streets or parking areas. Use self-closing doors to discourage the entrance of flying insects. Clearly mark the doors for *entrance* and *exit* to prevent accidents between those entering and leaving the facility.

Floors, Walls, and Ceilings

Proper construction, repair, and cleaning of floors, walls, and ceilings are important elements of an effective sanitation program. Consider the specific needs of food areas when making a selection of floors and walls, including:

▲ Sanitation

▲ Safety

▲ Durability

▲ Comfort

▲ Cost.

Smooth and easy-to-clean surfaces are needed in food preparation areas, store rooms (including dry storage areas and walk-in refrigerators), and warewashing areas.

For floor coverings, walls, wall coverings, and ceilings, use non-absorbent materials that are free from cracks or crevices where soil may lodge, and from imperfections that might cause accidents. Choose surfaces that are resistant to damage and deterioration from water, cleaning agents, and the repeated scrubbing used to keep them in good condition.

Floors

The preferred floor materials in food preparation and warewashing areas include:

▲ Terrazzo

▲ Quarry tile

▲ Asphalt tile

▲ Ceramic tile.

Concrete may also be used if it has been properly sealed with an epoxy or similar material to make it durable and non-absorbent.

In food production and warewashing areas avoid the use of:

▲ Wood

▲ Vinyl

▲ Carpeting.

These materials are not easy to clean and they tend to absorb water, soil, and other forms of contamination.

The *Food Code* also prohibits the use of carpeting in:

▲ Food preparation areas

▲ Walk-in refrigerators

▲ Warewashing areas

▲ Toilet room areas where handwashing lavatories, toilets, and urinals are located

▲ Refuse storage rooms or other areas subject to moisture.

Floors graded to drains are needed in food establishments where water-flush methods are use for cleaning. In addition, the floor and wall must be coved and sealed. **Coving** is a curved sealed edge between the floor and wall that eliminates sharp corners or gaps that would make cleaning difficult and ineffective (Figure 8.2). In food establishments where cleaning methods other than water flushing are used for cleaning floors, the floor and wall juncture must be coved with a gap of no more than 1/32 inch (1mm) between the floor and wall.

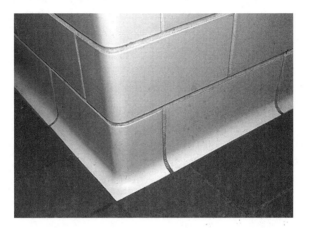

Figure 8.2—Coving

Slips and falls are the most common types of accidents in food establishments. In many instances, slippery floors cause accidents which result in personal injury and broken equipment. Use mats and other forms of anti-slip floor coverings where necessary to protect the safety of workers (Figure 8.3). These devices should also be impervious, non-absorbent, and easy-to-clean. The *Food Code* prohibits the use of sawdust, diatomaceous earth, or similar materials on floors except where they are applied in small amounts to collect liquid spills or drippage immediately before spot cleaning.

Figure 8.3—Anti-Slip Mat in Front of a Three-Compartment Sink

Walls and Ceilings

Smooth, nonabsorbent, and easy-to-clean walls and ceilings must be provided in food preparation and warewashing areas, walk-in refrigerators, and toilet facilities. Light colors enhance the artificial lighting in these areas and make soil easy to see for better cleaning. In areas that are cleaned frequently, use walls and wall coverings that are constructed of materials such as ceramic tile, stainless steel, or fiberglass. Concrete, porous blocks, or bricks should only be used in dry storage areas unless they are finished and sealed to provide a smooth, nonabsorbent, and easily cleanable surface.

Ceilings should be constructed of nonporous, easily cleanable materials. Studs, rafters, joists, or pipes must not be exposed in walk-in refrigeration units, food preparation and warewashing areas, and toilet rooms. Light fixtures, ventilation system components, wall-mounted fans, decorative items, and other attachments to walls and ceilings must be easy-to-clean and maintained in good repair.

Restroom Sanitation

Toilet facilities are required for all employees. Employee restrooms must be conveniently located and accessible to employees during all hours of operation. Toilet facilities near work areas promote good personal hygiene, reduce lost productivity, and permit closer supervision of employees. A toilet room on the premises must be completely enclosed and provided with a tight-fitting and self-closing door.

Materials used in the construction of toilet rooms and toilet fixtures must be durable and easily cleanable. The floors, walls and fixtures in toilet areas must be clean and well maintained. Supply toilet tissue at each toilet. Provide easy-to-clean containers for waste materials, and have at least one covered container in toilet rooms used by women.

Poor sanitation in toilet areas can spread disease. Dirty toilet facilities can also have a negative effect on the attitudes and work habits of the establishment's employees. Include these areas in the routine cleaning program to ensure they are kept clean and in good repair. Never store food in restroom areas.

Handwashing Facilities

In earlier chapters of this book, you learned about the importance of food workers' hands as sources of contamination and cross contamination. Food workers must know when and how to wash their hands in order to do their jobs safely. Conveniently located and properly equipped handwashing facilities are key factors in getting employees to wash their hands. Locate handwashing stations (Figure 8.4) in food preparation, food dispensing, and warewashing areas. Handwashing stations must also be located in or adjacent to restrooms. The number of handwashing stations required and their installation are usually set by local health or plumbing codes.

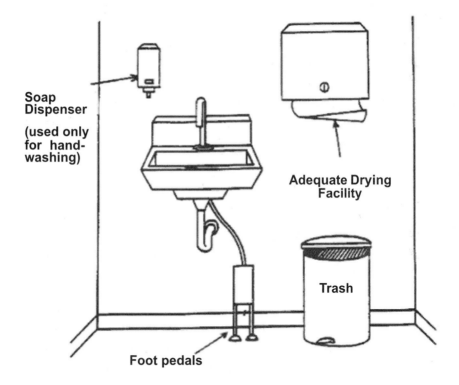

Soap Dispenser

(used only for hand-washing)

Adequate Drying Facility

Trash

Foot pedals

Figure 8.4—Handwashing Station
(Source: *FDA and CFP Food Establishment Plan Review Guide*)

A handwashing station must be equipped with hot and cold running water under pressure, a supply of soap, and a means to dry hands. Water at a temperature of at least 110°F (43°C) through a mixing valve or combination faucet is required. If a self-closing, slow-closing, or metering faucet is used, it must provide a flow of water for at least 15 seconds without the need to be reactivated.

Equip each handwashing station with a dispenser containing liquid or powdered soap. The use of bar soap is frequently discouraged by regulatory agencies because bar soap can become contaminated with germs and soil. Individual disposable towels and mechanical hot-air dryers are the preferred hand-drying devices.

Most local health departments do not recommend retractable cloth towel dispenser systems because there are too many possibilities for contamination. Common cloth towels, used multiple times by employees to dry their hands, are also prohibited.

Keep handwashing stations clean and well maintained and never use them for purposes other than handwashing. It is equally important to remember that sinks used for preparing produce and warewashing must **not** be used for handwashing.

Plumbing Hazards in Food Establishments

The *Food Code* includes many different components within the definition of a plumbing system. In particular, it identifies the

▲ Water supply and distribution pipes

▲ Plumbing fixtures and traps

▲ Soil, waste, and vent pipes

▲ Sanitary and storm sewers

▲ Building drains, including their respective connections and devices within the building and at the site.

A properly designed and installed plumbing system is extremely important to food sanitation. Contamination of the public water supply in a food establishment is a real public health problem. Numerous outbreaks of gastroenteritis, dysentery, typhoid fever, and chemical poisonings have been traced to cross connections and other types of plumbing hazards in food establishments.

Cross Connections

A **cross connection** is any physical link through which contaminants from drains, sewers, or waste pipes can enter a potable (safe to drink) water supply. A cross connection may be either direct or indirect.

A *direct cross connection* occurs when a potable water system is directly connected to a drain, sewer, non-potable water supply, or other source of contamination. An *indirect cross connection* is where the source of contamination (sewage, chemicals, etc.) may be blown across, sucked into, or diverted into a safe water supply.

One of the most common cross connections in the food industry is a garden hose attached to a service sink, with the end of the hose submerged in soapy water that is used to clean floors.

Backflow

Backflow is the backward flow of contaminated water into a potable water supply. It is caused by back pressure. It occurs under two conditions:

▲ Backpressure where contamination is forced into a potable water system through a connection that has a higher pressure than the water system

▲ Backsiphonage when there is reduced pressure or a vacuum formed in the water system (Figure 8.5). This might be caused by a water main break, the shut-down of a portion of the system for repairs, or heavy water use during a fire.

In the event of water pressure drop... backsiphonage occurs.

Figure 8.5—Backsiphonage

Methods and Devices to Prevent Backflow

The plumbing system in your food establishment must be designed, constructed, and installed according to the plumbing code in your jurisdiction. A properly designed and installed plumbing system will keep food, equipment, and utensils from becoming contaminated with disease-causing agents found in sewage and other pollutants. Cross connections and backflow can be prevented by using the following devices:

▲ Air gap

▲ Atmospheric vacuum breakers

▲ Pressure type vacuum breakers

▲ Double check valves

▲ Reduced pressure principle backflow preventers.

Air Gap

An **air gap** is the physical separation of the potable and non-potable system by a vertical air space. The vertical distance between the supply pipe and the flood rim must be at least two times the diameter of the supply pipe (2D), but never less than 1 inch (25mm). An air gap is the most dependable backflow prevention device. Figure 8.6 shows how an air gap may be used between a faucet and the rim of a sink and between a drain line and the house sewer system.

Atmospheric Type Vacuum Breaker

The most commonly used *atmospheric type anti-siphon vacuum breakers* incorporate an atmospheric vent in combination with a check valve. The valve's air inlet closes when the potable water flows in the normal direction. But, as water ceases to flow, the air inlet opens, thus interrupting the possible backsiphonage effect.

Atmospheric vacuum breakers can be used on most inlet type water connections which are not subject to back-pressure. This includes low inlet feeds to receptacles containing toxic and non-toxic substances.

Pressure Type Vacuum Breakers

Pressure type vacuum breakers are designed for use under continuous supply pressure, but are not good under backpressure. The device is usually spring loaded and designed to operate after extended periods of hydrostatic pressure. These devices should only be used where non-pressure vacuum breakers cannot be used.

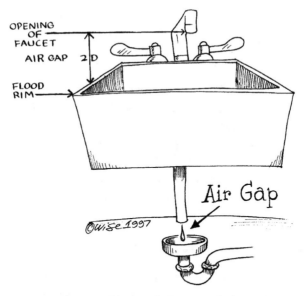

Figure 8.6—Air Gap

Double Check Valve

A *double check valve* consists of two internally loaded, independently operating check valves, together with tightly closing resilient seated shut-off values upstream and downstream of the check valves.

A double check valve assembly may be used as protection for all direct connections through which foreign materials might enter the potable water system.

Reduced Pressure Principle Backflow Preventers

This assembly consists of two internally loaded independently operating check valves and a mechanically independent, hydraulically dependent relief valve located between the check valves.

Reduced pressure zone devices may be used on all direct connections which may be subject to back-pressure or back-siphonage. Use this device when there is the possibility of contamination by material that is a potential health hazard.

These devices are commonly installed as main line protection to protect the municipal water supply. They should also be used on branch line applications where non-potable fluid would constitute a health hazard if introduced into the potable water supply system.

Grease Traps

Food establishments that do a lot of frying, charbroiling, and similar production processes will produce large amounts of grease. These facilities should be equipped with a grease trap that removes liquid grease and fats after they have hardened and become separated from the wastewater.

Grease traps are especially important when the food establishment is connected to a septic system or other type of on-site wastewater treatment and disposal system. A grease trap must be located for easily accessible cleaning. Failure to locate a grease trap so that it can be properly cleaned and maintained can result in the harborage of vermin and/or the failure of the sewage system.

Garbage and Refuse Sanitation

Proper disposal and storage of garbage and refuse protect food and equipment from contamination. **Good management of waste decreases attraction of insects, rodents, and other pests to the food establishment.** Typically, this involves proper handling and short-term storage of the materials inside the operation. Outdoor storage bins hold refuse and garbage for slightly longer periods until pickup. **Refuse** is solid waste which is not disposed of through the sewage disposal system. **Garbage** is the term applied to food waste that cannot be recycled.

An inside storage room and all containers must be large enough to hold any refuse, recyclables, and returnables that accumulate in the food establishment.

Provide a trash container in each area of the food establishment or premises where refuse is generated or commonly discarded, or where recyclables or returnables are placed. Do not put containers in locations where they might create a public health nuisance or interfere with the cleaning of nearby areas. Use durable, cleanable, insect and rodent-resistant, leak-proof, and non-absorbent equipment and receptacles to store refuse, garbage, recyclables, and returnables. Use plastic bags and wet strength paper bags to line receptacles for storage inside the food establishment or within closed outside receptacles. You must keep equipment and receptacles covered if they contain garbage and are not in continuous use or after they are filled.

Equipment and containers used for holding refuse, recyclables, and returnables must be cleaned regularly. Remember, dirty equipment and containers attract insects and rodents. When cleaning this equipment, be careful not to contaminate food, equipment, utensils, linens, or single-service and single-use articles.

Provide suitable cleaning equipment and supplies such as high pressure pumps, hot water, steam, and detergent to thoroughly clean equipment and receptacles. Waste water produced while cleaning the equipment and receptacles is considered to be sewage. It must be disposed of through an approved sanitary sewage system or other system that is constructed, maintained, and operated according to law.

Each food establishment should also have an outside storage area and enclosure (Figure 8.7) to hold refuse, recyclables, and returnables that accumulate. An outdoor storage surface should be durable, cleanable, and maintained in good repair. Keep equipment and receptacles covered with tight-fitting lids, doors, or covers to discourage insects and rodents.

Refuse and garbage should be removed from the premises as needed to prevent objectionable odors and other conditions that attract or harbor insects and rodents. Refuse removal may be required less frequently during winter months when the air temperature is colder.

Outdoor storage areas must be kept clean and free of litter. Suitable cleaning equipment and supplies must be available to clean the equipment and receptacles. Refuse storage equipment and receptacles must have drains, and drain plugs must be in place.

Figure 8.7—Outside Garbage and Refuse Station

Compactors and other equipment for refuse, recyclables, and returnables must be installed to minimize the accumulation of debris. Make sure that you clean under and around the unit to prevent insect and rodent harborage.

Cardboard or other packaging material (that does not contain food residues) may be stored outside for regularly scheduled pickup for recycling or disposal. If materials are stored so they do not create a rodent-harborage problem, a covered receptacle is not required.

Some food establishments may provide redeeming machines for recyclables or returnables. According to the *Food Code*, a redeeming machine may be located in the packaged food storage area or consumer area of a food establishment if food, equipment, utensils and linens, and single-service and single-use articles are not subject to contamination from the machines and a public health nuisance is not created.

Pest Control

All food establishments must have a pest control program. Insects and rodents which spread disease and damage food are the targets. These include rats, house mice, house flies, cockroaches, small moths, and beetles. Insects and rodents carry disease-causing bacteria in and on their bodies.

The benefits of proper cleaning and sanitizing of equipment and utensils, time and temperature controls, and food handling can all be wasted if insects and rodents are allowed to contaminate foods and food contact surfaces. Any food establishment may have an occasional insect or rodent problem. However, their continual presence causes major problems and indicates a lack of good sanitation and control measures.

The key element of a successful pest control program is prevention (Figure 8.8). However, no single measure will effectively prevent or control insects and rodents in food establishments. It takes a combination of three separate activities to keep pests in check. You must:

▲ Prevent entry of insects and rodents into the establishment.

▲ Eliminate food, water, and places where insects and rodents can hide.

▲ Implement an integrated pest management program to control insect and rodent pests that enter the establishment.

Figure 8.8—Prevent Pests from Invading Your Establishment

Insects

What insects lack in size, they more than make up for in numbers. Insects may spread diseases, contaminate food, destroy property, or be nuisances in food establishments. Insects need water, food, and a breeding place in order to survive. The best method of insect control is keeping them out of the establishment coupled with good sanitation and integrated pest management when needed.

Flies

House flies, blow-flies, and fruit flies are the types of flies most commonly found in food establishments.

The house fly is the one most likely to spread disease. Twenty-one species of flies are categorized as "disease-causing flies" because they are proven carriers of *E. coli*, *Salmonella*, *Shigella* and other germs that can cause foodborne illness (Olsen). When a fly walks over filth, such as human and animal feces or garbage, some of the material, including bacteria, sticks to its body and leg hairs. When a fly feeds on waste material, it takes some bacteria into its body. When the fly goes into an area where food is prepared or eaten, it walks on food and food contact surfaces thereby contaminating them (Figure 8.9). The house fly cannot chew solid food. Instead, the fly vomits on solid food to soften it before eating it. In vomiting, some of the bacteria are spread on food and food-contact surfaces and cause them to become contaminated.

Figure 8.9—Flies Spread Bacteria

Blow-flies are usually larger than house flies and are a shiny blue, green, or bronze color. Blow-flies have a keen sense of smell and are attracted by the odors produced by food establishments and food-processing plants.

Fruit flies are the smallest of the three flies and are attracted by decaying fruit. Fruit flies are known to spread plant diseases. Their role in the spread of germs that cause infections in humans is currently being investigated by researchers at the USDA.

Fly Control

Eliminate the insect's food supply as a first step to fly control. Store food properly to protect it from flies. In many cases the main source of flies is improperly stored garbage. Store garbage and other wastes in fly tight containers that are cleaned regularly.

Good sanitation includes proper cleaning of kitchen, dining, and toilet facilities and should be a routine practice.

Equip windows, entrances, and loading and unloading areas with tight fitting screens or air curtains (Figure 8.10) to prevent the entry of flying insects.

Figure 8.10—Air Curtain

Insect electrocuter traps are devices used to control flying insects such as moths and houseflies. The traps contain a light source that attracts the insects to a high voltage wire grid. Upon contact with the grids, the high voltage destroys the flies and other insects. Recent evidence suggests that insect fragments are produced as the insects are destroyed. These fragments can be a source of contamination for food and food-contact surfaces. The *Food Code* prohibits insect electrocutor traps from being installed over a food preparation area. When used in food establishments, these devices must be installed so that dead insects and insect fragments will not be propelled onto or fall on exposed food; clean equipment, utensils, and linens; and unwrapped single-service and single-use articles.

Chemical insecticides should be applied by a professional pest control operator. Remember, insecticides are a supplement to, never a substitute for, a clean food establishment.

Cockroaches

The four most common cockroaches in the United States are the American, Oriental, German, and Brown Banded. The German cockroach is the one most frequently found in food establishments (Figure 8.11).

Similar to flies, cockroaches are capable of carrying disease organisms on their body. They crawl from toilets and sewers into kitchens, running over utensils, food preparation areas, and unprotected food. They carry bacteria on their hairy legs and body as well as in their intestinal tract.

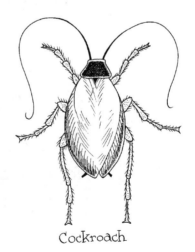

Cockroach

Figure 8.11—German Cockroach

Roaches avoid light and commonly hide in cracks and crevices under and behind equipment and facilities. It is possible to have a cockroach infestation and not realize it until they are caught by surprise some night when a light is turned on suddenly in a room.

Cockroach Control

To control cockroaches, maintain good housekeeping indoors and outside. Eliminate hiding places by picking up unwanted materials, such as boxes and rags. Also fill cracks and crevices in floors and walls and around equipment. Doors and windows should be tight fitting. Doors and windows, if kept open for ventilation or other purposes, must be protected by screening, air curtains, or other effective means.

Check incoming food and supplies for signs of infestation such as egg cases and live roaches. Store food in containers that are insect proof and have tight-fitting lids. Keep floors, tables, walls and equipment clean and free of food wastes. Frequent cleaning will help remove egg cases and reduce the roach population.

Chemical control is only recommended in combination with the other control procedures and not as the primary method. Insecticides are made to kill. It is recommended that insecticides be applied by a reputable, professional pest control operator. These individuals know how to handle insecticides safely to avoid contamination of food and food-contact surfaces.

Cockroaches are best controlled by insecticides that remain active for several days and are sprayed into cracks and crevices. Make certain the pest control operator remembers to treat baseboards, windows, door trim, and under and around appliances. Commercially available baits can be used in "blind" areas, such as false ceilings. Re-treatment in a few days may be necessary to kill any young roaches that may have recently hatched.

Moths and Beetles

Moths and beetles cause concern since they invade certain foods and can do extensive damage. The saw-toothed grain beetle, flour weevil, and rice weevil are examples of stored product beetles that can be found in food establishments. These insects feed on a variety of products including corn, rice, wheat, flour, beans, sugar, meal, and cereals. Small moths and beetles create problems of wasted food and nuisance rather than disease.

Control of Moths and Beetles in Stored Foods

Control of moths and beetles begins with proper stock rotation. Use the FIFO (first in, first out) system of stock rotation. All opened packages or sacks should be either used immediately or stored in covered containers. Clean shelves and floors frequently. These pests thrive on flour, meal, and cereal products that are spilled on the floor or shelving.

Examine incoming shipments for signs of infestation, and keep infested products away from other stock until they are ready for disposal. Keep dry food storage areas cool. Cool temperatures limit growth of these insects and reduce egg laying. Insecticides may have to be use to prevent re-infestation of stored food. Pest control operators must treat only along baseboards, cracks, crevices, and under pallets to avoid contaminating food.

Rodents

Rodents adapt easily to human environments and tolerate a wide range of conditions. They may carry germs that can cause a number of diseases including salmonellosis, plague, and murine typhus. Rodents also consume and damage large quantities of foods each year (Figure 8.12).

The term "domestic" rodents includes *Norway rats, roof rats*, and *house mice*. The Norway rat is the most commonly found rat in the United States. Common names for the Norway rat are the brown rat, sewer rat, and wharf rat.

Figure 8.12—Types of Rodents

The *Norway rat* is primarily a burrowing rat. It hides in burrows in the ground and around buildings and in sewers. Norway rats will eat almost any food but prefer garbage, meat, fish, and cereal.

They stay close to food and water, and their range of travel is usually no more than 100 to 150 feet.

The *roof rat* is smaller than the Norway rat but is a very agile climber. It generally harbors in the upper floors of buildings but is sometimes found in sewers. Roof rats prefer vegetables, fruits, cereal and grain for food. The range of travel for the roof rat is also about 100 to 150 feet.

The *house mouse* is the smallest of the domestic rodents. It is found primarily in and around buildings, nesting in walls, cabinets, and stored goods. The house mouse is a nibbler, and it prefers cereal and grain. Its range of travel is 10 to 30 feet.

Signs of Rodent Infestation

It is unusual to see rats or mice during the daytime, since they are nocturnal. It is necessary to look for signs of their activities. From rodent signs you can determine the type of rodent, whether it is a new or old problem and whether there is a light or heavy infestation.

Droppings

The presence of rat or mouse feces (Figure 8.13) is one of the best indications of an infestation. Fresh droppings are usually moist, soft and shiny, whereas old droppings become dry and hard. Norway-rat droppings are the largest and have rounded ends. They look a lot like black jelly beans. Roof-rat droppings are smaller and more regular in form. The house mouse's droppings are very small and pointed at each end.

Norway Rat
Droppings

Roof Rat
Droppings

House Mouse
Droppings

Figure 8.13—Rodent Droppings

Runways and Burrows

Rats stay in a limited area. They are very cautious and repeatedly use the same paths and trails. Outdoors in grass and weeds, you can see paths worn down that are 2 to 3 inches wide.

The Norway rat prefers to burrow for nesting and harborage. Burrows are found in earth banks, along walls, and under rubbish. Rat holes are about three inches in diameter whereas mouse holes are only about one inch in diameter. If a burrow is active, it will be free of cobwebs and dust. The presence of fresh food or freshly dug earth at the entrance of the burrow also indicates an active burrow.

Rubmarks

Rats prefer to stay close to walls where their highly sensitive whiskers can keep in contact with the wall. By using the same runs, their bodies rub against the wall or baseboard. The oil and filth from the rat's body are deposited and create a black mark called a rubmark. Mice do not leave rubmarks that are detectable, except when the infestation is heavy.

Gnawings

The incisor teeth of rats grow 4 to 6 inches a year. As a result, rats have to do some gnawing each day to keep their teeth short enough to use. Gnawings in wood are fresh if they are light colored and show well-defined teeth marks.

Tracks

Tracks may be observed anywhere along rat or mouse runs both indoors and outdoors. Dust in little-used rooms and in mud around puddles are good places to look for tracks. Rat tracks may be 1-1/2" long.

Miscellaneous Signs

Rodent urine stains can be seen with an ultraviolet light. Rats leave a different pattern than mice. Rat and mouse hairs may be found along walls, etc. When examined under a microscope they can be distinguished from other animal hairs.

Rodent Control

Effective rodent control begins with a building and grounds that will not provide a source of food, shelter, and breeding areas. The grounds around the food establishment should be free of litter, waste, refuse, uncut weeds, and grass. Unused equipment, boxes, crates, pallets and other materials should be neatly stored to eliminate places where pests might hide.

Get rid of all unwanted materials that may provide food and shelter for rodents. This means storing trash and garbage in approved-type containers with tight-fitting lids. Plastic liners help to keep cans clean, but they will not exclude rats and mice. Garbage must be removed frequently from the food establishment to cut off the food supply.

Rats can enter a building through holes as small as 1/2 inch. Mice can enter through even smaller holes. Buildings and foundations should be constructed to prevent rodent entry, and all entrances and loading and unloading areas should be equipped with self-closing doors and door flashings that will serve as suitable rodent barriers (Figure 8.14). Metal screens with holes no larger than 1/4 inch should be installed over all floor drains.

Figure 8.14—Keep the Building Sealed Tight to Prevent Rodent Entry

Traps are useful around food establishments where rodenticides are not permitted or are hazardous. Live traps can be used for collecting live rats. Check traps at least once every 24 hours. Killer or snap traps can also be used as part of a rodent-control program. When using these types of traps, place them at right angles to the wall along rodent runways. Glueboards are shallow trays that have a very sticky surface. They catch mice when the mouse's feet stick to the board when they walk on it.

Rodenticides are dangerous chemicals that can contaminate food and food contact surfaces if not handled properly. Baits should be used outdoors to stop rodents at the outer boundaries of your property. These substances should be applied by a professional pest control operator. Baits should be placed in a tamperproof, locked bait box that will prevent childrean and pets from being exposed to the toxic chemicals inside (Figure 8.15). Always make certain that pesticides are stored in properly labeled containers, away from food in a secure place. Dispose of containers safely and know emergency measures for treating accidental poisoning.

Figure 8.15—Bait Boxes and Glueboard
(Courtesty of J.T. Eaton & Company, Inc.)

Integrated Pest Management (IPM)

Modern pest control operators use integrated pest management to control pests in food establishments. **Integrated pest management** is a system that uses a combination of sanitation, mechanical, and chemical procedures to control pests. Chemical pesticides are used only as a last resort and only in the amount needed to support the other control measures in the IPM program.

The National Pest Control Association recommends a five-step program for IPM:

1. Inspection

2. Identification

3. Sanitation

4. Application of two or more pest management procedures

5. Evaluation of effectiveness through follow-up inspections.

There are many benefits of an integrated pest management program. It is cost effective and more efficient than programs using only chemicals. IPM is also longer lasting and safer to you, your employees, and your customers.

Summary

The case presented at the beginning of the chapter illustrates how pests can pose a very serious problem for food establishments. Pests contaminate food and food contact-surfaces, and they can carry disease agents that are harmful to humans. Pest are attracted to food establishments by the food, water and good odors they find there. In order to keep pests under control, food establishments must prevent them from entering the facility, eliminate sources of food and shelter, and implement an Integrated Pest Management (IPM) program.

Customer opinion surveys show that cleanliness is a top consideration when choosing a place to eat or shop for food. Customer satisfaction is highest in food establishments that are clean and bright and where quality food products are safely handled and displayed.

Proper construction and repair and cleaning of floors, walls, and ceilings are important parts of an effective sanitation program. Sanitation, safety, durability, comfort, and cost are the main criteria you will use when selecting materials for floors and walls. Surfaces of floors and walls should be resistant to damage and deterioration from the water, detergents, and repeated scrubbings used to keep them clean. Walls and ceilings should be light colored to show soil, and enhance the artificial lighting used in kitchen areas.

Good personal hygiene, which includes proper handwashing, is an important factor in preventing the spread of foodborne disease organisms. Handwashing stations must be conveniently located to allow food handlers to do the frequent hand cleaning required to do their job safely.

Clean and suitably equipped toilet facilities must be provided for employees. These facilities must be kept clean and in good repair to prevent the spread of disease and promote good personal hygiene.

A properly designed, constructed, and installed plumbing system is very important to food sanitation. Numerous outbreaks of gastroenteritis, dysentery, typhoid fever, and chemical poisonings have been traced to cross connections and other types of plumbing hazards in food establishments. Air gaps and mechanical vacuum breakers are used to protect the municipal water supply. Consult a professional plumber or your local plumbing code for details about the plumbing requirements in your jurisdiction.

Proper storage and disposal of garbage and refuse are necessary to prevent contamination of food and equipment and avoid attracting insects, rodents, and other pests to the food establishment. Proper facilities and receptacles must be provided inside and outside the establishment to hold refuse, recyclables, and returnables that may accumulate. Refuse and garbage should be removed from the premises frequently enough to minimize the development of objectionable odors and other conditions that attract or harbor insects, rodents, and other pests.

Case Study 8.1

John Smothers is in charge of remodeling the meat department at Kelly's Supermarket. The meat department provides a variety of meat, poultry, and fish items and specializes in customized orders. The equipment in the facility is not new but still meets current design and construction standards. However, the floors and walls in the department are old and need to be upgraded.

1. What criteria should John use when considering alternative floor and wall materials for the facility?

2. Why is wood flooring not recommended for this area?

3. What color wall materials are recommended? Why?

4. What features about a meat department should influence the selection of floor and wall materials?

Case Study 8.2

Ryan's Hamburgers is a small restaurant that specializes in hamburgers and tacos. A prep cook at Ryan's was recently diagnosed with Hepatitis A. As a result, several employees and customers were required to be immunized against the disease. An inspection by the local health department revealed that the handwashing lavatory in the food prep area at Ryan's was out of service. The only lavatories available to employees on duty were the ones in the restrooms.

1. Does the handwashing situation at Ryan's pose a health risk for customers? How?

2. What action should the local health department take against Ryan's?

3. What action should the owner/operator of Ryan's take to prevent a similar incident in the future?

4. What are some of the likely effects this episode will have on Ryan's restaurant?

Case 8.3

Scientists from North Carolina State University have demonstrated how easily cockroaches can spread *Salmonella* organisms among themselves and to foods. The research, reported in the February, 1994 issue of *Journal of Food Protection*, confirms that cockroaches are capable of acquiring and infecting other cockroaches and objects.

Cockroaches are attracted by warmth, darkness, food, and moisture. Given the number of ways cockroaches can become contaminated—by feeding on infected food or feces, contacting an infected cockroach directly or indirectly, or by drinking an infected water source—the risk they pose should be taken seriously. The fact that cockroaches can acquire and infect each other and other objects with *Salmonella* bacteria is an important point for the food industry to remember.

▲ What is the best way to control cockroaches in a food establishment?

Answers to the case studies are provided in Appendix A.

Discussion Questions (Short Answer)

1. What do customer opinion surveys show as one of the top reasons people choose a place to dine or shop for food?

2. How can the premises of a food establishment be used to make a positive impression on customers?

3. Identify four characteristics that floors, walls, and ceilings should have when they are used in kitchens and warewashing areas?

4. List the primary components of a handwashing station.

5. How does the location of handwashing facilities influence sanitation and personal hygiene?

6. How does proper refuse and garbage disposal contribute to the pest control activities of a food establishment?

Quiz 8.1 (Multiple Choice)

1. Which of the following statements about toilet facilities is **false**?

    a. Toilet facilities must be available for all employees.

    b. Employee toilet facilities must be conveniently located and accessible to employees during all hours of operation.

    c. Separate toilet facilities should be provided for men and women.

    d. Poor sanitation in toilet facilities will influence customers' opinions about cleanliness but will not promote the spread of disease.

2. The most effective device for protecting the potable water system from contamination by backflow is a (an):

    a. Air gap.

    b. Double check valve.

    c. Hose bib.

    d. Vacuum breaker.

3. When an air gap is used, the vertical distance between the supply pipe (faucet) and the flood rim must be at least:

   a. Two times the diameter of the supply pipe, but never less than 1/2 inch.

   b. Two times the diameter of the supply pipe, but never less than 1 inch.

   c. Three times the diameter of the supply pipe, but never less than 2 inches.

   d. Four times the diameter of the supply pipe, but never less than 3 inches.

4. Which of the following statements about garbage and refuse sanitation is **false**?

   a. Proper disposal and storage of garbage and refuse are necessary to prevent contamination of food and equipment and avoid attracting insects and rodents.

   b. A trash receptacle must be provided in each area of the establishment where refuse is generated.

   c. The equipment and receptacles used to store refuse, garbage, etc., must be durable, clean, non-absorbent, leakproof, and pest-proof.

   d. Trash may be stored outdoors in plastic bags provided the bags are stored at least 15 inches off the ground.

5. You are surveying your food establishment for unsafe conditions. Which one of the following situations requires **corrective action**?

   a. A trash receptacle with the lid off while in use.

   b. A handwashing station with a multi-use cloth towel for hand drying.

   c. Light colored ceramic tile being used for the walls of the food-preparation area.

   d. Anti-slip flooring provided in the warewashing area.

6. Backsiphonage is likely to occur if:

   a. The pressure in the potable water system drops below that of a non-potable or contaminated water source.

   b. Contamination is forced into a potable water system through a connection that has a higher pressure than the water system.

   c. Pressure builds up in a sewer line due to blockage.

   d. The water seal in a kitchen trap is siphoned out.

7. The primary responsibility of food establishment managers in pest control is to ensure:

   a. Good sanitation that will eliminate food, water, and harborage areas.

   b. Pesticides are applied safely.

   c. The pest control operator they use employs integrated pest management.

   d. The parking area is kept free of litter.

8. The best way to encourage employees to wash their hands when needed is to:

   a. Provide a separate restroom for employees and customers.

   b. Provide properly equipped handwashing stations convenient to work areas.

   c. Provide hand sanitizers instead of handwashing lavatories in food-preparation areas.

   d. Put up a sign in the employee locker room reminding them of the importance of proper handwashing.

9. Coving is a (an):

   a. Curved sealed edge between the floor and wall that eliminates sharp corners to make cleaning easier.

   b. Anti-slip floor covering used to protect workers from slips and falls.

   c. Plastic material used to seal cracks and crevices under and around equipment in a food establishment.

   d. A device used to prevent backsiphonage.

10. What factor has the greatest influence on where people choose to eat or shop for food?

    a. Cost of the food.

    b. Nutrition of food.

    c. Quality of service.

    d. Cleanliness of food and facilities.

Answers to the multiple choice questions are provided in Appendix A.

Suggested Reading/References

Cichy, R. F. (1994). *Quality Sanitation Management*. Educational Institute of the American Hotel and Motel Association. Store Publishing Company. East Lansing, MI.

Food and Drug Administration and Conference for Food Protection (1997). Food Establishment Plan Review Guide. Washington, DC.

Khan, M. A. (1991). *Foodservice Operations and Management*. Van Nostrand Reinhold. New York, NY.

Knight, J. B. and L.H. Kotschevar (1989). *Quantity Food Production*. Van Nostrand Reinhold. New York, NY.

Kopanic, R. J., B.W. Sheldon, and C.G. Wright (1994). "Cockroaches as Vectors of Salmonella: Laboratory and Field Trials," *Journal of Food Protection*, Vo. 57., No. 2, pp. 125-32, February, 1994.

Longrèe, K. and G. Armbruster (1996). *Quantity Food Sanitation*. Wiley. New York, NY.

Olsen, Alan R. (1998). "Regulatory Action Criteria for Filth and Other Extraneous Materials—Review of Flies and Foodborne Enteric Diseases," *Regulatory Toxicology and Pharmacology*, 28:199-211.

Pannell, D. V. (1990). *School Foodservice Management*. Van Nostrand Reinhold. New York, NY.

West, B. B., et al. (1986) *Food Service in Institutions*. Macmillan Publishers. New York, NY.

Selected Web Sites

Actron, Inc.
www.actroninc.com

Association of Applied Insect Ecologists
www.aaie.com

B&G Equipment
www.bgequip.com

Ecolab
www.ecolab.com

EPA Office of Prevention, Pesticides and Toxic Substances
www.epa.gov/internet/oppts

Hobart Corporation
www.hobartcorp.com

Insect-O-Cutor
www.insect-o-cutor.com

The Orkin Company
www.orkin.com

PCO
www.pco.ca

Pest Control Industry
www.pestweb.com

Plumbnet-Industry Resources
www.plumbnet.com/resources.html

Steritech
www.stertech.com

Do-It Yourself Pest Control
www.doyourownpestcontrol.com

Virginia Polytechnic Institue and University Pesticide Programs
www.vtpp.ext.vt.edu

NSF International
www.nsf.org

Underwriters Labratories Inc.
www.ul.com

National Restaurant Association
www.restaurant.org

The Food Marketing Institute
www.fmi.org

Gateway to Government Food Safety Information
www.foodsafety.gov

United States Department of Agriculture (USDA)
www.usda.gov

United States Foods and Drug Administration (FDA)
www.fda.gov

9 ACCIDENT PREVENTION AND CRISIS MANAGEMENT

Water Outage Can Leave Food Establishment High and Dry

On occasion, food establishments will experience a temporary loss of utility service such as power, water supply, or sewer service. These service interruptions can be particularly disturbing if they occur during peak periods of food production and service.

A disruption of water service can cause many problems for a food establishment. It makes food production difficult, and it makes proper cleaning and sanitizing of equipment, utensils and the general environment nearly impossible.

A food establishment should have a contingency plan in place to help it cope with unplanned events such as an interruption of utility service. Contingency plans are like insurance. While we hope we will never need them, we cannot afford to be without them.

Learning Objectives

After reading this chapter, you should be able to:

▲ Recognize how human error leads to accidents

▲ Discuss the importance of a safety audit

▲ Recognize the value of safety inservice training

▲ Identify public health rules and regulations that pertain to accidents and crisis management

▲ Develop a first aid plan

▲ Discuss the need for a fire exit plan, fire drill practice, and fire extinguisher use

▲ Identify the types and required locations of fire extinguishers

▲ Post a list of emergency phone numbers by the telephone

▲ Identify the responsibility to comply with Occupational Safety and Health Administration (OSHA) rules.

Essential Terms

Cardiopulmonary resuscitation (CPR)

Dead man's switch

Safety in Food Establishments

Food establishments contain many hazards that can cause accidents and injuries. Falls, cuts, scrapes, puncture wounds, and burns occur when food handlers become careless or equipment is not in good repair. The main point to remember is that accidents, like foodborne illnesses, can be prevented.

Accident prevention programs typically involve three types of activities:

▲ Eliminating hazards in the environment

▲ Providing personal protective equipment for workers

▲ Training workers about hazards and how to prevent or avoid them.

Hazards can be eliminated by putting safety devices on equipment, keeping equipment in proper working condition, and keeping the environment free of obstacles and potential hazards.

Built-in safety guides keep hands away from moving blades during grinding and slicing operations. Cuts can be prevented when using slicers and deboning equipment by wearing steel and nylon woven fiber gloves. Hot pads protect hands against burns. Work shoes with closed toes and rubber soles protect the worker's feet from injury.

Workers must be taught how to prevent and avoid hazards that cause injury. They must know such things as:

▲ When and how to use ladders for climbing

▲ How to lift things properly

▲ Why safety devices must be left in place on equipment

▲ How to dress for safety.

Before they start to work, give employees instructions on specific procedures and safety measures related to the assigned tasks. An ounce of prevention is worth far more than a pound of cure.

Each establishment needs a basic first aid kit (Figure 9.1) placed in a readily accessible area but away from all food, equipment, utensils, linens, and single-use articles. Depending on the size of the operation, more than one kit may be needed. The kit should include an instruction manual describing basic first aid treatments, sterile dressings, adhesive tape and bandages, burn cream, antiseptic ointment, and other basic first aid supplies. Check expiration dates on supplies monthly and replace them as needed. Post a list of important telephone numbers such as fire department, police, emergency medical system, and nearest hospital emergency room beside the telephone in a readily accessible area.

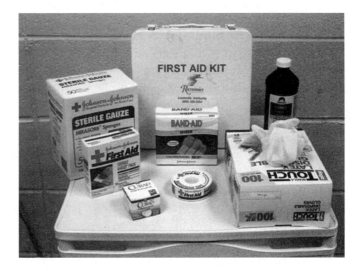

Figure 9.1—First Aid Kit

What to Do if an Accident Occurs

1. Stay calm. Information is needed to determine the seriousness of the injury. Begin by checking to see if the victim is responsive. Kneel and ask, "Are you OK?"

2. Examine the injury and decide whether outside help is needed. If in doubt, call for help. Victims who are not

breathing and do not have a heartbeat have a much greater chance of survival if they receive prompt medical care in a hospital or by trained paramedics.

3. Administer first aid according to the type of injury.

4. Keep unnecessary personnel away from the victim.

5. Record the victim's name, the date and time of accident, type of injury or illness, any treatment given, and amount of time it takes for emergency assistance to arrive.

Common Types of Injuries

Falls

Falls are preventable. Keep floors clean and free from debris and spills. Wet floors, grease, or spilled food are hazards. Use signs to alert others of newly mopped wet floors. Use anti-slip mats to cover areas that have high traffic and are frequently wet (see Figure 8.3). If someone falls, ask him or her to stay still, then determine the extent of injury. Seek medical assistance if there is a question about the victim's ability to move. Improper use of ladders or climbing on chairs, boxes, etc., instead of using a ladder can cause employee falls. Always use the proper equipment according to directions and intended use.

Cuts

Cuts are common injuries in food establishments. Carelessness with knives, slicers, choppers, grinders, mixers, and broken glass cause most problems. Keep knives sharp and store separately from other utensils. Accidents occur frequently when dull tools slip and cause an injury. Never leave knives on counters or submerged in a sink full of soapy water. Always use the proper tool for the task. Do not try to catch a falling knife. Let it fall to the floor. Do not try to clean the circular blade of a slicer while the equipment is still assembled. *Always make sure slicers and other cutting equipment are unplugged before taking them apart for cleaning.*

When treating a cut, first find out the extent of the wound. Then apply pressure to the site with a clean cloth or towel and elevate the injury. If the cut is minor, wash thoroughly with mild soap and water. If available, apply antibiotic ointment. Apply a clean, water-resistant bandage. If the person will return to food preparation duties, have them wear disposable gloves to provide an extra layer of protection. More severe cuts should be checked by a medical professional.

Always use power equipment according to the manufacturer's directions. Leave guard devices in place—they protect you from injury. A dead man's switch is commonly provided on large pieces of equipment, such as mixers, grinders, and cutters. A **dead man's switch** allows equipment to operate only when it is engaged. If the switch is released, the equipment will shut off immediately, thereby reducing the risk of injury. Usually the switches are located so that you have to stretch your arms to each side to maintain the pressure. Thus, you cannot accidentally grab for something in or on the machine without releasing contact with the switches and the machine stops. Overriding the switch with duct tape or other such material is illegal. Caution employees against such actions. The switch is there to protect them from serious harm.

Burns

Burns are classified as first degree (redness and pain), second degree (blisters, redness, and pain), or third degree (charring of skin layers, little or no pain). The extent of a burn is important. If a small area is burned, either first or second degree, first aid is usually all that is needed. Third degree burns must be checked by a medical professional. Burn treatment includes the following:

▲ Remove the source of the burn. If the burn is caused by an electrical source, rescuers must first determine if the electricity is off before touching the victim.

▲ Soak the burned area in cool water to soothe minor burns.

▲ Over-the-counter pain medications may be used to help relieve pain and reduce inflammation and swelling.

▲ Cover the burn with a sterile bandage or a clean dry dressing. Do not puncture blisters.

▲ Keep the victim calm and quiet.

If a victim receives a second- or third-degree burn or if the burn covers a large area, you should seek medical assistance immediately. Burns on the face, hands, or groin area are considered serious and should be checked by a medical professional.

Chemical burns require immediate treatment. Call for medical help and check the material safety data sheet (MSDS) for emergency interventions. Chemical burns are often caused by mishandling dishwashing compounds and cleaning solvents.

Poisoning

Poisoning can occur when food contaminated by chemical substances is eaten. *If there is any question that a food is contaminated by chemicals, do not wait, call for medical assistance immediately.* Poison Control Centers should be contacted, and you should follow instructions until help arrives.

Body Mechanics Classes

A body mechanics class can be used to teach workers how to lift, reach, and pull correctly in order to prevent back injuries. Back injuries are very expensive when you factor in lost productivity, workmen's compensation, and medical care. Insurance companies and many local health departments may provide pamphlets, videos, and class materials to promote back safety.

Employee Medications

According to the *Food Code,* only those medications necessary for the employee's health are allowed in a food establishment. This does not apply to medicines that are stored or displayed for retail sale. Employee medication must be clearly labeled and stored in an area away from food, equipment, utensils, linens, and single-use items like straws, eating utensils, and napkins. Medications stored in a food refrigerator must be kept in a clearly labeled and covered container located on the lowest shelf of the refrigerator.

Cardiopulmonary Resuscitation (CPR) and First Aid for Choking

Unless it is a requirement in your jurisdiction, you do not have to know how to administer first aid for choking or **Cardiopulmonary Resuscitation (CPR)**. However, posting charts with instructions for these procedures and other lifesaving steps can be helpful in emergencies (Figure 9.2). These charts are available from the American Red Cross and the American Heart Association. It is desirable to have at least one person on each shift trained in emergency response procedures and first aid. Remember, knowing how to do the right thing at the right moment can save a life.

Figure 9.2—Lifesaving Steps

(Source: American Red Cross - used with permission)

Local public health departments, American Red Cross, American Heart Association, National Safety Council, insurance companies, emergency medical systems, fire departments, and hospitals can provide classes, demonstrations, posters, videos, and other materials for use in food establishments. Personnel who volunteer to take safety related courses for use in the establishment are tremendous assets to the establishment.

Incident Reports

When an accident does occur, it is always advisable to fill out an incident report. Insurance companies require this documentation when processing claims. The time of the incident, conditions that contributed to the problem, extent of observed injury, any treatment provided at the establishment or a medical facility, and prescribed aftercare should be recorded. Notify insurance companies as soon as possible after an accident has occurred.

Safety Training Programs

Provide instructions on how to operate machinery and equipment in an easy-to-use format. Place a procedure book with detailed information on care and use of equipment in a central place. Post individual instruction sheets in the work area. Determine if employees can read signs and instructions. If you don't think they can, use picture diagrams to show steps for safe operation. In the case of foreign speaking persons, they may not be able to read English. Some may not be able to read their own language either. Merely posting instructions in their language does not mean the information has been received.

Hold demonstrations and discussions of procedures or proper use of equipment at the beginning of the shift or other appointed times when the majority of staff can attend. These mini-inservice sessions are extremely helpful and ensure that all personnel hear the same information. During the meeting, encourage employees to contribute to the discussion or share their views on how to improve operational flow. Managers and supervisors are responsible for teaching employees basic safety practices, and they must monitor work habits to ensure compliance. Employees who are careless in their work or use equipment improperly can cause harm to themselves and others.

Self-Inspection Safety Checks

Just as self-inspection programs help identify food safety problems in food establishments, safety audits can help prevent accidents. Conducting a safety audit can be one of the most cost-effective tasks you can perform. Fire and accident prevention can result in considerable cost savings.

Fire Safety

Fires are common problems in food establishments. Grease fires (in the plenum of the hood canopy) are the number one cause of fires in food establishments. Electrical fires are number two. Regular inspections of electrical equipment and prompt repair of faulty wiring or equipment can decrease this problem. Avoid overloading circuits and improper venting or grounding. Teach and monitor employees on the hazards of pulling electrical plugs out of sockets by the cord. They must grasp the plug head and then pull.

Keep hoods, ventilating fans, filters, and adjoining walls clean and clear of greasy buildup. Perform regular maintenance according to the manufacturer's directions.

In case of a fire emergency the first thing to do is get out of the building and **STAY ALIVE**. Be sure you evacuate the building and the fire department is called.

Fire Extinguishers

Portable fire extinguishers must be operable, easy to find, and employed according to type of fire (Figure 9.3).

Type of Fire Extinguisher	Material Burning	Most Common Type of Extinguisher Used
Type A	wood, paper, cloth	Pressurized water or ABC
Type B	grease, gasoline, solvents	CO_2 or B or BC
Type C	motors, switches, electrical	CO_2 or C

Figure 9.3—Types of Fire Extinguishers

Pressurized water is not recommended for food production areas because it might be used on a grease fire by mistake. Grease floats on water and would easily spread. Water on electrical fires could cause extensive damage, too. Carbon dioxide (CO_2) is recommended because it is safe and leaves no damaging residue on machinery or food contact surfaces. Any exposed food must always be discarded. ABC multipurpose extinguishers do leave a residue. Local fire departments can provide information and demonstrations on the proper type of fire safety equipment and how to use it.

Because employee's clothing can ignite when working around stoves, a safety blanket should be stored in an area where it can easily be reached if needed. Use the blanket to smother the flames and then call for help.

If you detect a fire that you cannot extinguish quickly, call for help. Then shut off exhaust fans, turn off equipment, close all doors and windows, and evacuate the area. Every food establishment needs a fire safety plan. Clearly mark routes of evacuation. Hold fire drills at periodic intervals to ensure that all employees know how and where to exit if the building is on fire. Battery powered emergency lights are required in most jurisdictions. They automatically provide light when power is interrupted.

Fire safety inspectors check fire extinguishers to certify them as operable. Portable fire extinguishers should be checked or serviced at least annually and marked with tags that specify the next due date for service. Fines can be levied for out-of-date equipment. If an extinguisher weighs less than 40 pounds, it can be mounted five feet from the ground. If it weighs more than 40 pounds, it may be stored no more than three and one-half feet off the ground (*Food Code*).

Hood Systems

Hood systems are usually installed over cooking equipment and dishwashers to remove vapors, heat, and smoke. Some hood canopies may have a built-in extinguishing agent. These devices can effectively extinguish a fire. Employees need to know how they work. Hood systems should be installed and periodically serviced by fire prevention professionals.

Sprinkler Systems

Sprinkler systems are often required by law in new or remodeled buildings. Supplies must be stored the regulated distance from sprinkler heads. If a sprinkler head is accidentally broken off, contact your sprinkler service immediately.

Crisis Management

A crisis can occur at any time. Besides natural disasters such as fire, flood, storms, or earthquakes, food establishments must deal with interference in normal operations that can literally cripple production. Other types of crises that can have a significant impact on retail food operations are power outages, interruption of water and sewer service, an outbreak of foodborne illness, media investigations and unexpected loss of personnel. How do you handle difficult situations? Develop a plan to handle the crisis with a problem-solving approach.

First, evaluate the extent of the problem. Check the facts and decide what possible outcomes can be expected. Are you going to need outside help? If you are working with a group of employees and supervisors, analyze their talents and use a team approach. **One spokesperson is better than having too many individuals giving out information.** Maintain control within your organization.

Set priorities according to the resources that exist. Remain calm. Look for positives. Follow established rules and regulations as set up in codes. Identify any outside resources that can help solve the problem. Keep a record of actions and communications in case it should be needed. In all situations, be honest.

Foodborne Illness Incident or Outbreak

If dealing with a suspected foodborne illness problem, cooperate with the customer and seek help from your local food regulatory agency. Work with the local regulatory agency to determine what may have caused the customer's illness. Establish the time the suspected food was eaten and, if a sample of the product is still available, preserve it for laboratory analysis. You may ask the customer to save samples of products eaten recently at home, too. Has medical attention been sought? Listen to everything the consumer says and look for clues that might explain the problem.

Do not try to diagnose the problem or belittle the consumer's complaint. Keep a record of the interview. Assure the consumer that you will get back to him or her after investigating the problem. Make sure that you do contact the customer again after facts are gathered. **Do not admit that your establishment is at fault—you have no proof of that until facts are checked.**

Check the Flow of Food

If, after checking the flow of food you find no evidence to support the consumer's claim, you may decide to keep the problem in-house. As long as only one person is affected, an outbreak is not suspected. However, if two or more persons report the same problem, consult outside resources on how to proceed. When preparing food samples for testing, it is wise to keep a sample in your freezer for cross validation of laboratory results. Public health officials, insurance agents, and, if needed, an attorney can help you manage the problem.

Formal Investigations

Remove suspect items from use. Health inspectors may want to conduct a HACCP inspection at the establishment. Employees may be required to have medical examinations. Closing a facility for a period of time to facilitate cleaning may be needed. Keep in mind that whatever must be done should be accomplished as quickly as possible. Cooperation to protect public health is the end goal. In the final analysis, it does not matter where the error occurred. It must be corrected and prevented from happening again.

Whatever the crisis may be, use a calm, systematic approach and evaluate actions as much as possible. Crisis management is not an easy task. It is always useful to have some kind of protocol to guide you through the event. Put together a team or committee consisting of management and employees to regularly brainstorm worst-case scenarios. Plan disaster drills to keep employees informed and trained in crisis management skills. Designate a spokesperson who is responsible for talking to the media, should they become involved. Keep a notebook with instructions on how to proceed in emergency situations.

Summary

A food establishment must be prepared to deal with unexpected events such as accidents, severe weather, and interruption of utility service. In the case presented at the beginning of the chapter, the food establishment is faced with a disruption of water supply to the establishment. This might have occurred due to a water main break, a fire in the vicinity, or some other cause. When faced with a water outage, the managers and employees of a food establishment must determine how to best cope with the situation. Under extreme conditions, the food establishment may have to be closed until service is restored. In less dire circumstances, less drastic measures may be required. The establishment may restrict sales to prepackaged and prepared foods. It may have to provide disposable plates, cups and utensils for their guests to use. In all circumstances, customer safety must be the top priority of the food establishment. Never make a decision that will put the safety and well-being of your guest in jeopardy.

Accident prevention programs are good investments. Use community resources to help build an effective plan. Insurance companies and vendors can supply training aids on accident prevention. Fire departments and emergency healthcare providers can provide demonstrations and advice that prepares you for accidents or other problems linked to safety. If you never need the plan, consider yourself lucky. However, be prepared.

Case Study 9.1

It was 1989, a beautiful fall day in San Francisco. The World Series was just getting underway in Candlestick Park; and then, the ground began to shake!

Across the city, a banquet was just getting started at the culinary arts school. As the ground shook and buildings swayed, guests, students, and staff evacuated the building as quickly as possible.

Meanwhile, in the dining room, a large ornate crystal chandelier was loosened from the ceiling and came crashing down, spraying fragments of glass across the area, including the buffet tables that were loaded with food. When the tremors stopped, the managers in charge of the food service were faced with immediate decisions concerning how to proceed after this catastrophic event. Because it was evening, the keys for supplies kept for emergencies were not available.

1. Make a list of the problems faced in this emergency and decide what should be done first, second, and so forth.

2. What community resources are needed to cope with this disaster?

3. How would you manage the food without electricity and a potable water supply?

Case Study 9.2

As a new employee was refilling the deep fat fryer, some excess cooking oil spilled onto the floor. Later that morning, the fry cook fell due to the slippery floor. After the fall, the cook was still in severe pain and was unable to return to work.

1. Why did this accident happen?

2. How can you help the cook?

3. How can you prevent future falls like this one?

Answers to the case study questions are provided in Appendix A.

Quiz 9.1 (Fill in the blanks)

1. Employees should wear _____ _____ when slicing or removing bones from meat.

2. Apply _____ to stop the flow of blood from a cut.

3. Employee medications that require refrigeration must be stored in a _____ and _____ container on the lowest shelf of the refrigerator.

4. If a consumer reports a foodborne illness from something eaten at your establishment, you should listen attentively; and if the suspect food item is still present, _____ it from use.

5. If an employee has a burned hand and arm that appear charred and they are not complaining of pain, it is a _____ _____ burn and should be looked at by a medical professional.

6. The fire inspector checks all _____ _____ for expiration dates and locations in the establishment.

7. List at least four sources that supply instructional assistance for health and safety training.

8. When someone falls in the establishment, your first action is to tell them to _____ _____.

9. If someone is choking, use the _____ _____ to help them breathe.

10. Post copies of _____ _____ in prominent locations to assist in evacuation during a fire.

Answers to all fill in the blank questions are provided in Appendix A.

References/Suggested Readings

Educational Foundation (1998). *Serve-Safe*. National Restaurant Association. Chicago, IL.

Food and Drug Administration (1999). *1999 Food Code*. United States Public Health Service. Washington, DC.

Rue, N., National Assessment Institute (1994). *Handbook for Safe Foodservice Management*. Prentice Hall. Englewood Cliffs, NJ.

Selected Web Sites

American College of Emergency Physicians
www.acep.org

American Red Cross
www.redcross.org

Christie Communications
www.christie.ab.ca

Health Answers Medical Reference Library
www.healthanswers.com

Injury Control Resource Information Network
www.injurycontrol.com

Medicine Net Company
www.medicinenet.com

National Safety Council
  www.nsc.org

Occupational Safety and Health Administration
  www.osha.gov

Safety Online
  www.safetyonline.net

Safety Smart Company
  www.safteytalks.com

Vermont Safety Information Resources, Inc.
  www.hazard.com

Gateway to Government Food Safety Information
  www.foodsafety.gov

10 EDUCATION AND TRAINING

A Bad Case of Training

As she started her second day of work in the kitchen at the Zesty Restaurant, Tammy hoped she would get a little more training than she did yesterday. She wanted to do a good job, but this was her first time working in the retail food industry.

It was another busy day and no one had time to teach her about safe food handling. She was assigned two tasks. One was preparing chicken pieces ready for the chef to use in tonight's baked chicken special; the other, to prepare salad ingredients including lettuce, celery, and raw carrots. These were simple tasks that she just knew she could do correctly. She brought the box of raw chicken pieces out of the cooler and placed them in a tub of water in the sink, just like she did at home. She carefully trimmed the pieces on a cutting board and returned the tray of chicken to the cooler. Tammy had just started to wipe off the board with dry cloth when she was called to help with set-ups in the dining room. An hour later she returned to start the salad preparation. She went straight to the cooler, selected the vegetables, washed them, and placed them on the same cutting board that she had used for the chicken.

Two days later, three employees who had eaten the salad Tammy had prepared became sick with watery diarrhea that lasted about 3 days. Six patrons, who had eaten salad on the same day as the employees had, reported to the Health Department that they had become ill after eating at the Zesty Restaurant. Investigation by the Health Department revealed that ill victims had eaten salad prepared by Tammy. During an interview, investigators found that Tammy had used the same cutting board for both the raw chicken and the salad ingredients. No one had taken the time to explain the need for cleaning and sanitizing cutting boards between use for raw chicken and salad ingredients. A little training would have made a big difference.

Learning Objectives

After reading this chapter, you should be able to:

▲ Define training and the important role it has in food service safety and sanitation

▲ Identify the types of training available and how to use them effectively

▲ Develop an action plan for training employees and measuring the effectiveness of their training

▲ Identify resources to aid in food safety and sanitation training.

Essential Terms

Needs assessment

Standards

Transfer of learning

Visual aids

Training

Webster's Dictionary defines training in the following ways:

Train (trayn) *v.* 1. To bring to a desired standard of efficiency or condition or behavior, etc., by instruction and practice. 2. To make proficient with specialized instruction and practice. 3. To focus or direct.

Training (tra'ning) *n.* 1. The act, process, or routine of one who trains.

For many years, managers in food establishments viewed training as a cost of doing business, a "necessary evil." Today, leaders understand that training is not a cost to the operation, it is a critical and required investment. Fortunately, as with any good investment, the outcomes make training well worth the effort.

It is important to understand that there is a significant difference between having knowledge of safety and sanitation and having the ability to train effectively on safety and sanitation. This chapter is included in our book to help managers explore aspects of training. As a retail food industry, managers are required to impact the bottom line—quality of the operations and guest satisfaction. The best way to ensure the safety and well-being of customers and employees is to know how to train.

In all food operations, it is crucial for any employee who handles food to be properly trained on all the equipment and on safety and sanitation procedures. The final report from the FDA Ad Hoc Committee on Training states that over 80% of foodborne illness can be traced to a employee not following proper procedures. The best examples of this are proper time and temperature control, good personal hygiene, and cross contamination control for preventing foodborne illness. Keep in mind that safety and sanitation procedures should be integrated into each job performance and standard operating practice where food is involved. Each individual should be targeted for training in food safety and sanitation training.

Now that you have a starting point on training topics, consider the why's and how's of training.

Benefits of Training

There are many benefits to proper employee training including:

▲ **Improved Customer Satisfaction**

A well-trained staff provides the service and quality that customers expect. Satisfied customers are repeat customers who help make the business successful. Customers want good service and good food served at the right temperature and that is free of potential hazards.

▲ **Lower Turnover**

Key factors in employee satisfaction and job retention are opportunities to be successful and grow in skills and abilities. Most employees want to be trained to do their job effectively. Lower turnover means dollar savings to the organization, better quality of life for all of the employees in the organization, and higher client satisfaction from service provided from a well-trained, professional staff.

▲ **Lower Costs**

   Well-trained employees make fewer mistakes than do new employees, which lowers overall costs to the operation.

▲ **Fewer Accidents**

   Studies have shown that the number of accidents decreases with experience and good training. A staff trained on safety procedures will have the skills to handle minor emergencies more effectively.

▲ **Better Quality**

   Training can result in higher quality of product and a more efficient staff. This translates into higher customer and employee satisfaction and a better working environment.

Figure 10.1—Good Training

Barriers to Training

With the benefits of training being so great, why do so many food establishments resist educating their workers? Knowing the benefits of great training isn't enough. The key is in execution: Having a plan and then executing the plan. Some of the problems that keep training from being used effectively include:

▲ **Perceived Lack of Time**

Because many operations are extremely busy, it seems as though there is not enough time to train. You need to change your thinking. Look at how much time the training will save if everyone knows what is expected and does it right the first time.

▲ **Lack of Commitment**

Some managers just never set aside the time to train employees. Having a commitment to training your employees in safe food handling techniques is essential for the overall success of your operation.

▲ **Lack of Training Skill**

Many managers want to train but do not know how. The following pages will provide you with a guide to getting started.

▲ **Language/Literacy**

The retail food industry is noted for a diverse work population. For many employees, English is their second language. This creates communication barriers when trying to train. It is also not uncommon for us to employ people who are illiterate in both English and their own language. To overcome these barriers, create your training modules using graphics, symbols, and pictures in place of words. Show employees what you want rather than just telling them. Then have them demonstrate to you what they have learned. This will show you if they have really learned the task.

▲ **Lack of Self Discipline**

Training others challenges the manager's ability to organize and present learning opportunities to staff members. It takes self-discipline and commitment of time to accomplish this task.

A well thought-out training plan, executed consistently by a self-disciplined leader that involves the whole team, can break down all the barriers and provide a safe and prosperous food operation.

The Training Plan

The training plan contains the following criteria:

▲ Establish the Standards of Employee Performance

▲ Prepare to Train Employees on the Standards

▲ Conduct the Training on the Standards

▲ Measure Performance Against the Standards

Establish the Standards of Employee Performance

Standards are the procedures, policies and processes that are necessary if a food operation is to be successful. First, identify the standards you will use. In corporations, there are usually several places to find established standards. Resources include job descriptions for specific jobs within the operation, company policy or best practice manuals, HACCP guidelines, state and local laws on safe food handling and food safety, etc. In addition, identify standards you want set within your operation. Once you have identified the standards that all employees should practice, the next step is to determine how best to communicate and train all employees to meet these goals.

Prepare to Train Employees on the Standards

After you have identified the standards to be met by each employee, you need to consider the following questions:

▲ What are you going to teach?

▲ Who is going to do the training?

▲ How are you going to conduct the training, i.e., what method of training are you going to use?

▲ When are you going to conduct training and how long will it last?

▲ Where are you going to conduct the training, and what materials are you going to need?

What Are You Going to Teach?

What you teach depends upon several issues. If the employee is new, the types of things that you are teaching will be different from those for a seasoned employee. When determining what training topics should be presented, you will want to do a **needs assessment**. A needs assessment is a type of questionnaire or survey used to determine specific training topics or issues that need to be addressed in a given department, unit, or establishment. Look at your team and the standards of performance you have established. What areas are they doing well in and what areas have you recognized as being in need of improvement? Listen to your customers. Complaints from customers are good indicators that training is needed in the operation. Some standards to consider for training include:

▲ Emergency situations, including fire safety

▲ Safe food handling procedures

▲ Common causes of foodborne illness

▲ Proper use and cleaning of food equipment and utensils

▲ AIDS awareness

▲ Customer relations and workplace etiquette

▲ Company policies and procedures.

Contact state/local regulatory agencies and industry associations for assistance in setting up and delivering a food safety education program.

When you are designing training, it is important to break down tasks and concepts into small, workable components. For example, if you want to teach a new employee about food storage and you try to teach them about frozen storage, cold storage, and dry storage all at once, it could be confusing to the employee. It is more effective if you take each area separately and teach short sessions with a focused topic. One key principle in adult learning is keeping the training sessions short (30-60 minutes.)   Another key principle is that adult learners want to participate in their learning. They do not just want to sit back and watch. They learn best through participation. If you try to teach too much in a short time, they will not have time to assimilate the information.

It is also important to have a long-term plan for training. If you are training employees on food storage and you teach one segment on frozen storage today, you should have a plan for when you will teach them about cold storage and dry storage. Supervisors need to know what the employees have learned before they assign new tasks.

Who Is Going to Do the Training?

Once you select the topic, the next step is to determine who should do the training. For example, if you are training on personal hygiene, you may want to conduct that session yourself. When you train on chemical safety, you may want to contact you local supplier and have a trained representative conduct the training session for your employees.

Managers, supervisors, and other employees can conduct training, but local government agencies, suppliers, etc., will typically conduct special classes on specific subjects, often at no cost. It is wise to review their information and style of training delivery before they deliver the session to make sure the training will be appropriate for your employees.

Be certain that the person conducting the training session is the right person for the topic and the audience. Trainers should know who is attending the session, the current level of understanding of the employees, and their current skill levels.

What Method of Training Are You Going to Use?

Another key to establishing effective training in your operation is in determining the right method of training for the subject and audience. There are many different training methods that can be used. Some of the more common types include:

▲ **On-The-Job Training**

Typically, on-the-job training is done one-on-one at the place where the work takes place. It is typically conducted by a trained co-worker. Many times when a competent employee is assigned the responsibility of training another, his or her workload remains the same, and this can cause incomplete, rushed, or fragmented training. Do not penalize a successful worker with additional tasks while expecting the same level of performance. Teaching another, when it is done right, is difficult work.

One way this type of training can be very successful is through a train-the-trainer program. Identify star employees in each area of the operation, then train them on effective training techniques. Once they have been trained, make them responsible for presenting the training for their area. Frequent formal and informal communication between managers and trainers is crucial for success.

▲ Classroom Style Training

This group training method is good to use when everyone in a department is learning a new method or standard. It is also used when training is used to refocus the team and encourage teamwork. It is not effective, however, when information is very technical in nature or includes many steps. Straight lecture is usually the poorest way to train employees, but is often used because you can reach more people at one time. Have employees evaluate the training to determine its effectiveness.

▲ Individual Developmental Training

This type of training is focused on the learner who takes the initiative from his or her own ability to perform on the job. It is designed to be done over a period of time. For example, the trainee is given a list of questions or case studies and a list of resources. They are then instructed to research the questions using the resources to find the answers. This works very well with adult learners. Another principle of adult learning is that people learn in different ways at varying speeds. Individual Development Training allows them to work at their own speed and in their own way. It is important, when using this method, to check in with the employee at consistent, timely intervals (i.e., daily, weekly, etc.) to determine their progress.

▲ Computer-Based Training

In today's high tech world, computer-based training is becoming increasingly popular. There are many training programs and modules you can purchase to train employees. Computer-based training works best for "soft-skills" topics where you do not need a hands-on style, such as customer service, sanitation, foodborne illnesses, etc. It is important to remember that this method may not work for everyone. Many workers lack the skills to use computer-assisted learning. If you decide to use this approach, the Internet is a good place to start researching possible programs.

Remember, adults learn best with a variety of training materials regardless of the training method. Use videos, audiotapes, posters, pictures, demonstrations, workbooks, or any other method of presenting information that can enhance learning.

When Are You Going to Conduct Training, and How Long Will it Last?

The timing of training is sometimes the difference between effective training and ineffective training. Food operations are very busy and identifying a time to present material is difficult. Some things to considered when scheduling training include:

▲ What are the peak busy times in the operation? It would be unwise to schedule training first thing in the morning when the operation does a very brisk breakfast or early morning shopping business.

▲ Which days are the busiest days? Typically in food establishments, there are certain days of the week that are busier than others (this is not necessarily the case in institutional food service operations.) Try to schedule training for times that are not busy.

▲ Can I do the training in smaller groups or at different times to keep from interrupting the operation? Often the greatest obstacle in scheduling training is trying to get everyone together at one time. A possible solution is to break the group into two smaller groups.

Whenever you decide to do your training, it is important to schedule and post training sessions in advance. This allows employees to mentally prepare for training. It also sends the message that training is important and valued. Cancellation of scheduled classes should only happen in emergency situations.

How Long Should Your Sessions Last?

Keep your audience and their attention span in mind at all times. Whenever possible, keep your sessions to one hour or less. It is better to conduct frequent regular training, in short sessions.

Where Are You Going to Conduct the Training and What Materials Do You Need?

Once you have decided when your training is going to be, the next step is to determine where you are going to conduct it and what materials you are going to need. For example, if you were doing a class where you were reviewing and reinforcing the basics of sanitary food practices, you would probably do it in a meeting room or empty dining room, using a workbook or handbook that has the practices in it. A Study Guide has been developed to accompany this textbook. You may obtain a copy by calling 1-800-282-0693 and asking for the *Essentials of Food Safety and Sanitation Study Guide*. If however, you were conducting a training session on proper handwashing procedures, you would want to do it in smaller groups, around a sink. You would demonstrate proper procedures and then allow your participants to try it. The only materials you need may be a handout or the proper handwashing procedure poster displayed above the sink.

There is no right or wrong place to conduct training. The location should be determined by what you are trying to teach. Try to arrange for:

▲ Adequate lighting

▲ A comfortable temperature

▲ As little noise as possible

▲ A site where everyone can see and hear what is being taught.

Conduct the Training on the Standards

Now that the preparation is completed, the next step is to conduct the training. Points in a great training session include:

▲ **Keep it Simple**

One common mistake in conducting training is trying to cram too much information into one session. In the end, the learner is overwhelmed and frustrated. When designing your training, focus on the main points you are trying to share with employees. Don't stray from your focus. For example, when teaching someone the proper handwashing technique, stick to the steps and present them in order. Don't start talking about grooming and uniforms standards as well. That could be another short training module or discussion.

▲ **Make it Fun**

Training should be an enjoyable experience for both the trainer and trainee. When learning is fun, the participants remember more of what was taught. Use colored pictures and get the trainees involved as much as possible. Include a team activity or energizer in your training session. Simple games and small prizes can help make the class fun and will reinforce learning.

▲ Involve the Audience

Employees want to be involved in the learning. Asking open-ended questions, having small group discussions, playing games, having quizzes, and including them in demonstrations are all ways of getting your participants involved. Use pre- and post-tests to evaluate the outcomes of the training.

▲ **Use Visual Aids**

We are a highly visual society. Television has taught us the value of pictures and diagrams to make a point. To enhance learning in food settings, it is important to use visual aids in training. Videos, charts, posters, models, slides, etc., are effective **visual aids**. A PowerPoint visual aid package has been developed to accompany the *Essentials of Food Safety and Sanitation* training program. You may obtain a copy by calling 765-342-8320. Make sure any visual aids are large enough for your trainees to see. If you use a flip chart and write too small, it can frustrate the trainee because they can't see what's going on. Keep in mind that great training is a combination of a skilled trainer and appropriate visual aids. Use visual aids as an enhancement, not a replacement for instruction.

Measure Performance Against the Standards

Once the training session is completed, it does not mean training is over. Whether the goal of the training session was to learn a new skill or to reinforce established practices, you want to be sure that the **transfer of learning** has taken place. Transfer of learning has taken place when an employee changes his or her behaviors based on the training. Proper follow up after the training session can help to reinforce the message in the training. Giving employees a test or having them demonstrate a procedure immediately following the training session can determine if the employees understood the information presented in the session.

To make sure that learning has taken place, consider the following actions:

▲ Provide follow-up discussions and reminders when the employee is back on the job to help reinforce the learning that took place in the training session.

▲ Observe whether or not people are doing the job correctly. Correct performance and changing behavior are the keys to improved food operations. Watch your employees perform their jobs after the training is completed. Look for the new skills they have learned. When they use their new skill, praise them. When they do not, coach them on the right way.

▲ If an employee is having trouble with a new skill, retrain him or her.

▲ Conduct post-training written tests or observe employees using a checklist developed using the criteria of the training class. Use written tests with caution. Employees who are not literate need special consideration. Verbal questions work better than written tests. Once again, recognize outstanding performance. Also follow up verbal recognition with a hard copy in the employee's record.

Be positive and consistent with your feedback. Be a good communicator. Provide verbal and written feedback to the employee. Ask the employee for suggestions that would help in learning the task. Finally, treat everyone with respect. Ridicule and sarcasm do not belong in training sessions.

Training Moments

In addition to scheduled training sessions, remember that training occurs all day long in food operations. Through manager and trainer actions, employees learn proper behavior. Be a positive role model for your employees. Use training moments to improve performance. When employees are not practicing proper procedures, be a coach. Teach them the right way to do it. Reinforce good habits by recognizing the employee who does it right. Look for learning readiness. Adults must see relevance in what they are asked to learn. One they accept the reason for training, they usually remember it well.

Summary

Remember our story at the beginning of the chapter? Proper training could have prevented the victims from getting sick. Tammy should have been trained about cross-contamination and the proper food safety practices that she could use to prevent it. Tammy must understand that raw products, such as chicken, frequently contain bacteria and other microbes that cause foodborne illness. These germs are transferred from a raw food to a cooked and ready-to-eat food during cross contamination. Tammy should have been taught when and how to clean and sanitize the cutting board and other utensils she had used to prepare the raw chicken before she used them to prepare the vegetables for the salad.

Training is an essential part of all retail food operations. With proper training, you can enjoy improved customer satisfaction, lower turnover, lower operating costs, fewer accidents, and a better quality staff. Establishments where training is not valued may see the results in poor employee performance, accidents, incidents of foodborne illnesses, and loss of business.

Creating a training plan is the first step in providing a quality training experience. Consider the following elements when putting together a training plan:

1. Establish the standards of employee performance you expect. Use all the resources available to you to help ensure you are focused on the "right" things

2. Prepare your training. Focus on the what, who, how, when, and where of training

3. Conduct your training session by keeping it simple, making it fun, involving your trainees, and using visual aids

4. Measure trainee performance based on the standards you taught them. Measure for both short-term understanding and long-term success

5. Remember that in addition to the formal training, you are teaching people all day long though your actions and behaviors. Be a good role model for your employees.

Case Study 10.1

The U-Go market decided to expand their business by opening a delicatessen section to serve fried chicken, potato salad, macaroni salad, and cold cuts of meat. After the equipment arrived, the vendor offered to give employees training on how to safely operate the equipment. At first, the owner refused, saying, "Training takes time and time is money." The certified food manager in charge of this area objected, "The products to be sold are all potentially hazardous foods."

1. Why did he say that?

2. What would you do in this situation?

Answers to the case study questions are provided in Appendix A.

Discussion Questions (Short Answer)

How many "wrongs" can you find in this picture?

Quiz 10.1 (Multiple Choice)

1. Today's food industry leaders understand that training is:

    a.   A cost with benefits.

    b.   Not necessary.

    c.   An investment.

    d.   A one-time event.

2. One of the benefits of effective, consistent training is:

    a. Higher turnover.

    b. Fewer accidents.

    c. Better quality.

    d. B and c.

3. The first step in creating a training plan is:

    a. Establish the standards of employee performance.

    b. Conduct training.

    c. Provide training moments.

    d. Evaluate the outcome of training.

4. What, who, how, when, and where are questions that need to be answered in which of the following?

    a. Deciding on what training materials are needed.

    b. Preparing to train employees.

    c. Evaluating training activities.

    d. All of the above.

5. Location of training is primarily based on:

    a. Transfer of learning.

    b. Employee attitudes.

    c. What type/subject you are trying to communicate.

    d. Who is doing the training.

6. A goal(s) in a great training session are:

    a. Make it difficult.

    b. Make it fun.

    c. Don't use visual aids.

    d. Make it passive.

7. To ensure the transfer of learning, one thing to consider is:

    a. Providing follow-up discussions.

    b. Catching people doing it wrong.

    c. Serving refreshments in class.

    d. Provide training at the end of the shift.

8. Informal role modeling/coaching training is known as

    a. The Discipline Process.

    b. Group training meetings.

    c. Training moments.

    d. Using visual aids.

Answers to the multiple choice questions are provided in Appendix A.

References/Suggested Readings

Esque, T. J., and T. F. Gilbert (1995). "Making Competencies Pay Off." *Training*. January. Vol. 32, No.1, pp. 44-50.

Felix, C. (1996). *Food Protection Report*. Charles Felix Publications. Leesburg, VA.

Longrèe, K., and G. Armbuster (1996). *Quantity Food Sanitation* Wiley and Sons. New York, NY.

Odiorne, G. S., and G. A. Rummler (1988). *Training and Development: A Guide For Professionals*. Commerce Clearing House, Inc. Chicago, IL.

Rue, N. (1994). *National Assessment Institute Handbook for Safe Foodservice Management*. Prentice Hall. Englewood Cliffs, NJ.

Selected Web Sites

Gateway to Government Food Safety Information
  www.foodsafety.gov

United States Department of Agriculture (USDA)
  www.usda.gov

Centers for Disease Control and Prevention (CDC)
  www.cdc.gov

United States Food and Drug Administration (FDA)
  www.fda.gov

USDA Food Safety and Inspection Service (FSIS)
  www.fsis.usda.gov

USDA/FDA Food and Nutrition Information Center
  www.nal.usda.gov/fnic/

United States Environmental Protection Agency (EPA)
www.epa.gov

Partnership for Food Safety Education
www.fightbac.org

The Food Marketing Institute
www.fmi.org

National Restaurant Association
www.restaurant.org

Prentice Hall
www.prenhall.com

11 FOOD SAFETY REGULATIONS

Voluntary Recall of Macaroni and Cheese Issued

A mid-west company recently issued a voluntary recall of 7.25 oz. boxes of Macaroni and Cheese identified with the code numbers FEB2500C21, FEB2600A21, MAR0400B21, and MAR0400C21. The product was being recalled because eight customers in northern California reported finding small pieces of glass in the products.

The affected macaroni and cheese product was distributed to supermarkets in northern California, western Nevada, Hawaii, Colorado, Wyoming, Nebraska, New Mexico, South Dakota, Virginia, Maryland, and the District of Columbia. Stores that received the macaroni and cheese product with the above codes have been notified to remove the product from their shelves and not offer it for sale to customers.

The manufacturer urges anyone who has already purchased the product not to eat it. Customers who wish to obtain a refund should discard the macaroni and cheese packets and return the box to their local grocery store.

Learning Objectives

After reading this chapter, you should be able to:

▲ Recognize the role of federal, state, and local jurisdictions in regulating and monitoring food safety

▲ Identify aspects of food operations that are included in inspections and audits

▲ Utilize this chapter as a reference for federal, state, and local authorities

▲ Identify organizations related to the food industry.

Essential Terms

Adulterated

Conference for Food Protection (CFP)

Generally Recognized as Safe (GRAS) substances

Grading

Inspection for Wholesomeness

Misbranded

Recall of a food product

State and local Departments of Agriculture

State and local Departments of Health

Environmental Protection Agency (EPA)

Food and Drug Administration (FDA)

United States Department of Agriculture (USDA)

State and Local Regulations

State and local agencies that are responsible for enforcing food and safety requirements may be under the department of environmental health, public health, or of agriculture. Jurisdictions varies from state to state. Contact your local authorities to find out who is in charge of food safety in your area.

Most food establishments will work closely with their local health department. This agency can provide you with a copy of its current food safety code. **Your state and local code contains the food safety standards and regulations that would apply to your type of operation**. Familiarize yourself with the key food safety provisions of the code.

Permit to Operate

Operating a food establishment usually begins when you obtain a permit from the local authority to operate. You cannot operate a food establishment unless you have a valid permit issued by the regulatory authority in your area. Apply for the permit well in advance of the anticipated date for opening the food establishment. Many health agencies require submission and approval of blueprints and specifications for an establishment before new construction or extensive remodeling can begin. Contact your local health agency to find out what level of plan review is required.

When a permit is obtained, post it in a prominent location where it can be seen. Permits are generally not transferable from one person to another, from one food establishment to another, or from one food operation to another. If you add a partner or incorporate, these actions are considered ownership changes and may necessitate application for a new permit to operate.

The minimum number of inspections or audits that must be performed each year is usually prescribed in the retail food code. However, many health departments are currently using a risk-based approach for setting inspection frequencies. Under this system, the risk a food establishment poses is calculated using such factors as the overall sanitation history of the establishment and number of:

▲ Meals served

▲ Potentially hazardous foods on the menu

▲ Critical violations observed.

Low-risk facilities are usually inspected once a year or less. High-risk facilities, on the other hand, may be inspected four or more times per year. A risk-based inspection system enables local health departments to target their scarce resources at those establishments that need them most.

Always cooperate with health department personnel. Greet the inspector when he or she arrives and ask to see official identification. Whenever possible, accompany the inspector during an inspection or audit. A routine inspection or audit normally consists of three phases:

Phase I is a pre-inspection conference where the environmental health specialist or sanitarian and the person in charge review previous inspection results and discuss information that is relevant to the current inspection.

Phase II is the inspection itself.

Phase III is the post-inspection conference where the results of the current inspection are reviewed and discussed.

Most inspections will focus on:

▲ Foods and supplies

▲ Personal hygiene and employee health

▲ Temperatures of food and food-holding equipment

▲ Cleaning and sanitizing procedures

▲ Equipment and utensils

▲ Water supply and waste disposal

▲ Pest control and other aspects of the operation that might compromise food safety.

If a critical violation such as temperature abuse or cross contamination is discovered during an inspection, corrective action must be taken immediately. Failure to do so could result in fines and possible closure of the establishment. Lesser violations should be corrected by the next inspection or audit.

Food managers should periodically conduct self-inspections of their establishment to ensure proper sanitation and food safety between health department inspections. During a self-inspection, the manager will look for signs of contamination or conditions that might be hazardous to food safety.

Federal Agencies

The primary federal agencies that protect our food supply include the:

▲ Food and Drug Administration (FDA)

▲ United States Department of Agriculture (USDA)

▲ Environmental Protection Agency (EPA).

Other important government agencies include the:

▲ National Marine Fisheries Service (NMFS)

▲ Centers for Disease Control and Prevention (CDC)

▲ Occupational Safety and Health Administration (OSHA)

▲ Consumer Product Safety Commission (CPSC)

▲ Federal Trade Commission (FTC).

Please note that states have their own counterparts of agriculture, health, OSHA, and EPA.

Food and Drug Administration (FDA)

The Food and Drug Administration (FDA) is an agency of the United States Department of Health and Human Services. The FDA is recognized as the leading food and drug regulatory agency in the world. This agency regulates the processing, manufacturing and interstate sale of many food items. All foods must comply with FDA food safety regulations EXCEPT domestic or wild cattle, sheep, swine, goat, horse and mule; and domestic chicken, turkey, duck, goose, guinea fowl, and egg products.

FDA seeks to protect the public's health by preventing food adulteration and misbranding. **Adulterated** food is food that contains filth, is decomposed or otherwise unfit, is produced in unsanitary conditions, or contains poisons or harmful substances damaging to health. **Misbranded** food is falsely or misleadingly packaged or labeled. It may be represented as a specific food product or contain ingredients not included on the label but does not meet national standards for that food. Misbranded food may also be packaged in such as way as to be misleading. For instance, it may be an imitation of another food, but the label does not notify the consumer that it is an imitation.

FDA's goal is to ensure that food shipped in interstate commerce is safe, pure, wholesome, sanitary, and honestly packaged and labeled. The agency also helps to prevent milk-borne and shellfish diseases and assists in the control of sanitation related to shellfish production.

The FDA also sets standards with respect to composition, quality, labeling, and safety of foods and food additives. FDA's duties include performing routine audits of food processing plants, food storage, and imported foods to ensure that safety standards are maintained. The agency is also responsible for the toxicological safety of food packaging materials.

The FDA has published the *Food Code* (Figure 11.1) which serves as a model for retail food programs that are sponsored by federal, state, local, and tribal agencies. The *Food Code* is not a federal law or federal regulation, and it does not replace existing retail food laws. The Code is more like a set of recommendations put forth to

Figure 11.1—Food Code

promote food safety and sanitation nationwide. The *Food Code* is law (statute, regulation, or ordinance) in all jurisdictions that have adopted it—either by reference or section-by-section.

The FDA works closely with other federal, state, local and tribal food regulatory authorities in matters of food protection. There is also close cooperation between the FDA and the food processing industry in their concern for the cleanliness and wholesomeness of the nation's food supply.

The FDA can request a recall of a food product if the agency has determined that a potential hazard exists in a particular food product which would justify regulatory action should such action become necessary. Food recalls are "voluntary" with the possible exception of a recall involving infant formula. In the vast majority of cases, food processors voluntarily remove products from the marketplace to keep consumers safe. However, if a firm does not agree with an FDA request for a voluntary recall, the agency may issue publicity and proceed to seize the product in order to remove it from commerce.

Regulators and industry personnel agree that a properly managed voluntary recall is the most rapid, effective, and efficient means of achieving product removal. Three types of food recalls, Class I, Class II, or Class III, can be issued depending on the severity of the health risk (Figure 11.2). It is important that all food retail establishments have a policy regarding the handling of food recalls. Recalled foods should be promptly removed from retail shelves and cannot be sold.

Class I	Foods that may cause serious adverse health consequences
Class II	Foods that would result in a temporary or reversible health problem
Class III	Foods that are not likely to cause danger to health

Figure II.2—Food Recalls

U.S. Department of Agriculture (USDA)

The **United States Department of Agriculture (USDA)** has many agencies engaged in food protection. The Agricultural Marketing Service (AMS) conducts programs that aid in the marketing of agricultural products. The AMS develops quality standards and offers voluntary grading services for red meats, poultry and eggs, dairy products, and fruits and vegetables.

Many types of quality standards have been developed to help ensure food quality and define differences in quality that affect the usefulness and value of a food. Grade or quality standards cover the whole range of a product's natural quality. The number of grades for a particular food depend on its variability. For instance, eight grades are necessary to cover the range of beef quality, whereas only three are needed for poultry. Beef grades usually encountered are USDA Prime, Choice, and Good. Choice is the grade most widely available; it denotes a high degree of tenderness, juiciness, and flavor. USDA Grade A poultry has more meat and a better appearance. Lower grades, such as B and C; are seldom offered for retail sale, but are used in processed foods such as soup, and pot pies.

The Food Safety and Inspection Service (FSIS) of the USDA inspects domestic and imported meats, poultry, and processed meat and poultry products. Employees of the FSIS inspect meat in accordance with the *Federal Meat Inspection Act of 1906*, to ensure a clean, wholesome, disease-free meat supply. Inspections are made by veterinarians or persons under their supervision at places of animal slaughter and at meat processing facilities. Poultry is inspected under the provisions of the *Federal Poultry Products Inspection Act of 1957*. This law requires poultry to be processed in plants having full-time government inspection service. Poultry is inspected live prior to slaughter, during evisceration (removal of the internal organs), and during or after packaging. The *Egg Products Inspection Act of 1970* provides for the inspection of egg product processing plants. It also provides for a surveillance program to help ensure that eggs still in the shell and that are unfit for human consumption (incubator rejects, leakers, etc.) are being properly restricted or disposed of by egg handlers.

Purchasers should be aware of the difference between inspection and grading. **Inspection for wholesomeness** refers to an official examination of food to determine whether it is wholesome and free from adulteration. **Grading** refers to the process of evaluating foods relative to specific, defined standards in order to assess its quality. Inspection is often required by law, but grading is optional. Most purchasers prefer to buy graded products because they know those products have met specific quality standards. Of the two services, inspection for wholesomeness is the most important for food safety.

U.S. Department of Commerce (USDC)

Grade standards for processed fishery products have been established by the Voluntary Inspection Service of the U.S. Department of Commerce. The Department's National Marine Fisheries Service (NMFS) administers these programs on a voluntary, fee-for-service basis. A fish processor may decide to have an inspector stationed in his plant on a permanent basis to assess the premises for sanitary conditions, determine that wholesome ingredients are used, and grade the product. This type of inspection is strictly voluntary. The product gets evaluated and graded as A, B, or C in quality, the last grade being the lowest acceptable quality. Examples of NMFS-regulated products include frozen raw and fried breaded fish portions, shrimp, frozen fried scallops, and frozen sole and haddock fillets.

Environmental Protection Agency (EPA)

The **Environmental Protection Agency (EPA)** was established in 1970. It was created to prevent, control, and reduce air, land, and water pollution. Related to food safety, the EPA regulates the use of toxic substances (pesticides, sanitizers, and other chemicals), monitors compliance, and provides technical assistance to states.

The EPA regulates pesticides under two major federal statutes. Under the Federal Insecticide, Fungicide, and Rodenticide Act (FIFRA), the EPA registers pesticides for use in the United States and prescribes labeling and other regulatory requirements to prevent unreasonable adverse effects on health or the environment. Under the Federal Food, Drug, and Cosmetic Act (FFDCA), the EPA establishes tolerances (maximum legally permissible levels) for pesticide residues in food. Tolerances are enforced by the Food and Drug Administration (FDA) for most foods, and by the U.S. Department of Agriculture/Food Safety and Inspection Service (USDA/FSIS) for meat, poultry, and some egg products.

In 1996, Congress unanimously passed landmark pesticide food safety legislation known as the Food Quality Protection Act (FQPA). The FQPA mandates a single, health-based standard for all pesticides in all foods; provides special protections for infants and children; expedites approval of safer pesticides; and requires periodic re-evaluation of pesticide registrations and tolerances to ensure that the scientific data supporting pesticide registrations remain up to date.

The EPA is also mainly responsible for the implementation of ten major environmental laws. These laws cover an enormous range of environmental problems including air pollution, water pollution, hazardous wastes, noise pollution, and contaminated drinking water. Some of these laws, especially those controlling the handling and disposal of solid and hazardous wastes, may affect food establishment operations. The store manager should contact the state and local environmental agencies to be sure their store is in compliance with current regulations.

Centers for Disease Control and Prevention (CDC)

The **Centers for Disease Control and Prevention** (CDC) is the federal agency that is responsible for protecting public health through the prevention and control of diseases. The CDC supports foodborne disease investigation and prepares annual summaries and statistics on outbreaks of foodborne diseases including all those transmitted through food and water.

Occupational Safety and Health Administration (OSHA)

The **Occupational Safety and Health Administration (OSHA)** was created in the Department of Labor to enforce the Occupational Safety and Health Act. It enforces health standards and regulations on safety, noise, and other workplace related hazards to protect workers from illness and injury when they are on the job. The agency conducts investigations and inspections to assess health and safety conditions in places of employment.

Federal Trade Commission (FTC)

The **Federal Trade Commission (FTC)** is involved in enforcing various laws regarding marketing practices and national advertising of foods and other products.

Other National Food Safety Related Organizations

The Conference for Food Protection (CFP) is a non-profit organization that represents the major stakeholders in retail food safety. Its membership includes representatives of the food industry, government, academic community, professional organizations, and consumer groups. The mission of the CFP is to promote food safety and consumer protection. It meets at least every other year to identify food safety problems, make recommendations, and implement practices to ensure food safety. The CFP has no formal regulatory authority, yet over the years it has been able to effectively influence the content of model retail food laws and regulations.

Other organizations also help to protect our food supply. Many of the following organizations have good information and training materials:

▲ The American Public Health Association (APHA)

▲ The Association of Food and Drug Officials (AFDO)

▲ The Council on Hotel, Restaurant and Institutional Educators (CHRIE)

▲ The Food Marketing Institute (FMI)

▲ The Frozen Food Industry Coordinating Committee (FFICC)

▲ The Food Research Institute (FRI)

▲ Institute of Food Technologists (IFT)

▲ National Conference on Interstate Milk Shipments (NCIMS)

▲ International Association for Food Protection [formerly the International Association of Milk, Food, and Environmental Sanitarians (IAMFES)]

▲ The National Environmental Health Association (NEHA)

▲ The National Pest Control Association (NPCA)

▲ The National Restaurant Association (NRA)

▲ NSF International (NSF)

▲ The National Shellfish Sanitation Program (NSSP)

▲ Underwriters Laboratories Inc. (UL).

Food Law

Federal Food, Drug, and Cosmetic (FD&C) Act

The FDA administers the federal *Food, Drug, and Cosmetic Act* which was originally passed by Congress in 1938 to ensure that foods are safe, pure, wholesome, and produced under sanitary conditions. Several amendments have been added to the law since then. Among the most notable amendments are the *Miller Pesticide Amendment of 1954*, the *Food Additives Amendment of 1958*, and the *Color Additives Amendment of 1960*.

The *Miller Pesticide Amendment* establishes a procedure for the setting of safe amounts (tolerances) for residues of pesticides which may remain on fresh fruits, vegetables, and other raw agricultural commodities when shipped in interstate commerce.

The *1958 Food Additives Amendment* has a two-fold purpose: (1) to protect the public health by requiring proof of safety before a substance may be added to food, and (2) to advance food technology and improve the food supply by permitting the use of substances which are safe at the levels of intended use. Before something is approved for use by the FDA, data must be supplied proving the substance is free from harm. If approved, the FDA will then define limits that can be used in foods. Currently, there is a list of over 3000 approved food additives. Excluded are those food additives which have been designated as **GRAS** or *generally* *recognized as safe* substances. These are additives that have been used in foods for years and with no apparent ill effects. Substances that typically fall into the GRAS category are spices, natural seasonings, flavoring materials, fruit and beverage acids, and baking powder chemicals. The purpose of establishing a GRAS list is to recognize the safety of basic substances without the requirement for rigorous safety testing.

The *Color Additive Amendments of 1960* were enacted to control the use of natural and synthetic color additives. Color additives are substances added only for the color they give. Color additives must be safe under normal conditions of use and must not cause cancer when ingested by humans.

1906—Federal Meat Inspection Act

This law provides for the mandatory inspection of animals, slaughtering conditions, and meat processing facilities. It helps to ensure that meat and products containing meat are clean, wholesome, free from disease, and properly represented. Such products are stamped "U.S. Inspected and Passed by Department of Agriculture." This stamp is necessary for interstate commerce and imported goods. This law is enforced by the Food Safety Inspection Service of the Department of Agriculture.

For products not crossing state lines, the Wholesome Meat Act is enforced (usually governed by state health departments).

1957—Federal Poultry Products Inspection Act

This law provides for the mandatory inspection of poultry, slaughtering conditions, and poultry processing facilities. It helps to ensure that poultry and products containing poultry are clean, wholesome, free from disease, and properly represented.

1970—Egg Products Inspection Act

Under the provisions of this Act, the Food Safety and Inspection Service (FSIS) inspects facilities that process egg products. These are facilities that break, pasteurize, freeze, dry, and package egg products. The Egg Products Inspection Act also authorizes FSIS or its designee to provide a surveillance program to help ensure that eggs still in the shell and that are unfit for human consumption (incubator rejects, leakers, etc.) are being properly restricted or disposed of by egg handlers.

1993—Nutrition Labeling and Education Act

The USDA and the FDA share responsibility for enforcing the Nutrition Labeling and Education Act. This law provides information on specific nutritional guidelines and defines what needs to be on a Nutrition Label (Fig. 11.3). This label, which provides information on protein, fat, carbohydrate, and mineral content, helps consumers make informed choices when selecting foods.

Meltaway/French Mint Combo 5 1/2 oz.

Nutrition Facts

Serving Size: 42.0g, 5 pieces
Serving Per Container: 4

Amount Per Serving

Calories 248   Calories from Fat 150

	% Daily Value*
Total Fat 17g	26%
Saturated Fat 10g	50%
Cholesterol Less then 5mg	0%
Sodium 50mg	2%
Total Carbohydrate 22g	7%
Dietary Fibers 1g	4%
Sugars 20g	
Protein 2g	

Vitamin A 0%	Vitamin C 0%
Calcium 6%	Iron 2%

* Percent Daily Values are based on a 2,000 calorie diet. Your daily values may be higher or lower depending on your calorie needs:

	Calories	2,000	2,500
Total Fat	Less than	65g	60g
Sat Fat	Less than	20g	25g
Cholesterol	Less than	300mg	300mg
Sodium	Less than	2,400mg	2,400mg
Total Carbohydrate		300g	375g
Dietary Fiber		25g	30g

19361

Figure II.3—The Nutrition Label

(Source: www.nal.usda.gov/fnic/)

Recent Initiatives in Food Safety

National Food Safety Initiative

From farm to table, the federal government's National Food Safety Initiative is working to reduce the incidence of foodborne illness by strengthening and improving food safety practices and policies. The initiative seeks to:

▲ Expand education efforts aimed at improving food handling in homes and retail outlets

▲ Improve inspections and expand monitoring efforts

▲ Increase research to develop new and more rapid methods to detect foodborne pathogens and to develop preventive techniques

▲ Improve intergovernmental communications and coordination of responses to foodborne outbreaks. FDA, CDC, USDA, and EPA will form a new intergovernmental group to improve federal, state, and local responses to outbreaks of foodborne illnesses.

FoodNet

CDCs chief role in the Food Safety Initiative has been to enhance surveillance and investigation of infections that are usually foodborne. This is the primary mission of FoodNet, which is a collaborative effort among the Centers for Disease Control and Prevention (CDC), the Food Safety and Inspection Service (FSIS) of the United States Department of Agriculture (USDA), the Food and Drug Administration (FDA), and the eight Emerging Infections Program (EIP) sites.

FoodNet was created in 1995 to produce more accurate national estimates of the number of cases and sources of specific foodborne diseases in the United States. The primary objectives of FoodNet are to:

▲ Determine the frequency and severity of foodborne disease

▲ Determine the proportion of common foodborne diseases that result from eating specific foods

▲ Describe the epidemiology of new and emerging bacterial, parasitic, and viral foodborne pathogens.

FoodNet uses an active surveillance program and epidemiological studies to monitor the burden of foodborne diseases over time. FoodNet has also been used to document the effectiveness of new food safety initiatives in improving the safety of our food and the public's health.

PulseNet

PulseNet is a national network of public health laboratories that perform DNA "fingerprinting" on bacteria that may be transmitted by food. The network permits rapid comparison of these "fingerprint" patterns through an electronic database at CDC. Laboratories participating in PulseNet perform DNA "fingerprinting" on disease-causing bacteria isolated from humans and from suspected food, using standardized equipment and methods. Once patterns are generated, they are entered into an electronic database of DNA "fingerprints" at the state or local health department and transmitted to CDC where they are filed in a central computer.

CDC created PulseNet so that scientists at public health laboratories throughout the country could rapidly compare the "fingerprint" patterns of bacteria isolated from ill persons and determine whether they are similar. Similar fingerprint patterns suggest that the bacteria isolated from ill persons come from a common source, for example, a widely distributed contaminated food product. Strains isolated from food products by regulatory agencies can also be compared with those isolated from ill persons. Identifying these connections can help to detect outbreaks and remove contaminated foods from the marketplace.

The Fight BAC!™ Campaign

The Fight Bac!™ Campaign is an educational program developed by industry, government, and consumer groups to teach consumers about safe food handling. The focal point of the campaign is BAC (short for bacteria) a character that is used to make consumers aware of the invisible germs that cause foodborne illness (Figure 11.4). Even though consumers cannot see, smell, or taste BAC, it and millions more germs like it are in and on food and food contact surfaces. The Fight Bac!™ Campaign uses media and community outreach programs to teach consumers how to keep food safe from harmful bacteria through four simple steps: wash hands and surfaces often, prevent cross contamination, cook foods to proper temperature, and refrigerate foods quickly. The partners in the Fight Bac!™ Campaign use newspaper articles, public service announcements (PSAs), printed brochures, and school curricula to teach consumers how to lower their risk of foodborne illness.

Figure II.4—Fight Bac!™ Logo

Consumer Advisory

Food establishments that sell or serve raw or undercooked animal foods or ingredients for human consumption must inform their customers about the increased risk associated with eating these kinds of foods. The food establishment can satisfy the consumer advisory requirements by providing both a disclosure and a reminder. The disclosure is used to make consumers aware of the dangers of raw or undercooked foods, and of items that either contain or may contain raw or undercooked ingredients. The reminder is a notice about the relationship between thorough cooking and food safety.

Disclosure is satisfied when items are described such as oysters on the half-shell (raw oysters), and raw-egg Caesar salad; or items are asterisked to a footnote that states that the items:

▲ Are served raw, or undercooked; or

▲ Contain (or may contain) raw or undercooked ingredients.

The reminder is satisfied when the items requiring disclosure are asterisked to a footnote that states:

▼ Regarding the safety of these items, written information is available upon request;

▼ Consuming raw or undercooked meats, poultry, seafood, shellfish, or eggs may increase your risk of foodborne illness;

▼ Consuming raw or undercooked meats, poultry, seafood, shellfish, or eggs may increase your risk of foodborne illness, especially if you have certain medical conditions.

Disclosure of raw or undercooked animal-derived foods or ingredients and reminders about the risk of consuming such foods should appear at the point where the consumer selects the food. Both the disclosure and the reminder need to accompany the information from which the consumer makes a selection. That information could appear in many forms such as a menu, a placard listing of available choices, or a table tent.

Consumer advisories may be tailored to be product-specific if a food establishment either has a limited menu or offers only certain animal-derived foods in a raw or undercooked, ready-to-eat form. For example, a raw bar serving molluscan shellfish on the half shell, but no other raw or undercooked animal food, could elect to confine its consumer advisory to shellfish.

Summary

The case presented at the beginning of the chapter provides an illustration of a voluntary recall. Whenever a company suspects that something may be wrong with one of its products, it will issue a recall to ensure that the product is removed from store shelves and that customers who may have already purchased the product are warned not to eat it. Regulatory agencies will frequently participate in recalls to ensure that suspect food products are removed from sale as quickly as possible. This is just one example of the partnerships that have been formed between the food industry and regulatory agencies to ensure the safety of our food supply.

Federal, state, and local food safety regulations are important for maintaining a safe food supply. State and local health departments are critical when it comes to monitoring and enforcing each state or local food code. They evaluate food establishments to help ensure delivery of safe food to consumers.

Food establishments and regulatory agencies should work together to protect the health of customers and meet their expectations for food quality and safety.

Case Study 11.1

A few hours after lunch, a customer returns to a restaurant and complains to a waiter that he got sick after eating the chicken salad. The customer claims that the chicken salad he had eaten for lunch was responsible for his illness.

▲ How should the waiter handle this customer complaint?

Case Study 11.2

Inspector Jones arrives at Jon's food establishment for the quarterly inspection. Jon is really busy and tells Mr. Jones to go ahead. When Mr. Jones returns with the result of the inspection, Jon is surprised. Mr. Jones has noted that the sanitizer dispenser in the chemical dishwasher is broken.

▲ What should Jon do?

Answers to the case study and all multiple choice questions are provided in Appendix A.

Quiz 11.1 (Multiple Choice)

1. Before opening a food establishment and serving food to the public, which of the following must be done?

   a. Pass a food certification exam

   b. Acquire a Permit to Operate

   c. Attend 16 hours of training in food safety

   d. All of the above.

2. Which federal agency prepared the model *Food Code*?

   a. Centers for Disease Control and Prevention

   b. Environmental Protection Agency

   c. Food and Drug Administration

   d. Occupational Safety and Health Administration.

3. Which federal agency is responsible for ensuring the safety of domestic meat, poultry, and egg products?

   a. Centers for Disease Control and Prevention

   b. Environmental Protection Agency

   c. Food and Drug Administration

   d. United States Department of Agriculture.

4. Which federal agency is responsible for ensuring the safety of all foods **except** domestic meat, poultry, and egg products?

   a. Centers for Disease Control and Prevention

   b. Environmental Protection Agency

   c. Food and Drug Administration

   d. United States Department of Agriculture.

5. Which federal agency is responsible for regulations on safety, noise, and other working conditions?

   a. Centers for Disease Control and Prevention

   b. Environmental Protection Agency

   c. Food and Drug Administration

   d. Occupational Safety and Health Administration.

6. If the FDA determines that the ingestion of a food may "cause serious adverse health consequences," a voluntary food recall may be requested. This would be an example of what kind of recall?

   a. Class I

   b. Class II

   c. Class III

   d. Class IV.

7. Which legal document does the FDA abide by to protect the food supply?

   a. Federal Food, Drug, and Cosmetic Act

   b. 1906 - Federal Meat Inspection Act

   c. 1957 - Federal Poultry Products Inspection Act

   d. None of the above.

8. During an inspection of a food establishment, the manager should:

   a. Ask to see credentials

   b. Assist the inspector whenever possible

   c. Discuss any violations

   d. All of the above.

9. When a suspected foodborne illness claim is reported to a wait-person, what should be the first course of action?

   a. Remove all customers from the restaurant immediately

   b. Call the health inspector

   c. Locate the food manager and ensure that the suspect food is not served

   d. Contact CDC.

10. Responsibilities of a food manager include:

    a. Obtaining copies, understanding, and complying with rules and regulations regarding the facility

    b. Handling food establishment self-inspections

    c. Handling claims of foodborne illness

    d. All of the above.

Answers to the multiple choice questions are provided in Appendix A.

References/Suggested Reading

IFT Publication (1992). "Government Regulation of Food Safety: Interaction of Scientific and Societal Forces." *Food Technology*. 46(1).

Potter, N.N., J. H. Hotchkiss (1995). *Food Science*. Chapman and Hall. New York, NY.

Labuza, T. P., and Baisier (1992). "The Role of the Federal Government in Food Safety." *Critical Reviews in Science and Nutrition*. 31(3). 167-176.

Selected Web Sites

Conference for Food Protection
www.uark.edu:80/misc/cfpncims/CFP1.html

Food and Agriculture Organization
www.fao.org

Gateway to Government Food Safety Information
www.foodsafety.gov

United States Department of Agriculture (USDA)
www.usda.gov

Centers for Disease Control and Prevention (CDC)
www.cdc.gov

United States Environmental Protection Agency (EPA)
www.epa.gov

United States Food and Drug Administration (FDA)
www.fda.gov

USDA Foods Safety and Inspection Service (FSIS)
www.fsis.usda.gov

USDA/FDA Food and Nutrition Information Center
www.nal.usda.gov/fnic/

National Registry of Food Safety Professionals
www.nrfsp.com

Occupational Safety and Health Administration (OSHA)
www.osha.gov

Partnerships for Food Safety Education
www.fightbac.org

World Health Organization
www.who.int

APPENDIX A

Answers to Case Studies and Quizzes

Chapter 1

Case Study 1.1

In this case study, you learned that *Salmonella enteritidis* is a common cause of foodborne illness. The bacteria were provided conditions that enable them to grow. First, the pies were only cooled for two and one-half hours after baking. The internal temperatures did not reach acceptable cold storage temperatures of 41°F (5°C) or below. Added to that, the pies were "temperature abused" (not refrigerated) during transport and during and after the outing. This provided optimal time and temperature, and the *Salmonella enteritidis* bacteria grew very rapidly under these conditions resulting in high levels of contamination.

Quiz 1.1	Quiz 1.2
1. b	1. False
2. d	2. False
3. b	3. True
4. a	4. True
5. c	5. True

Chapter 2

Case Study 2.1

Hazard: The foodborne hazard associated with this illness was likely a toxin produced by the bacteria *Staphylococcus aureus*. This organism may have originated from the food worker and was then transferred to the cheese by cross contamination. Being held at room temperature, the organism had the opportunity to grow in the cheese and produce a heat stable toxin. Even after cooking in the 500° F oven, the toxin was not destroyed. People that complained of illness probably had an intoxication due to *Staphylococcus aureus*. Also, cheddar cheese is drier and salty which provides ideal conditions for *Staphylococcus aureus* while inhibiting other microorganisms that might otherwise compete with the *Staph.* germs.

Prevention: There are three important prevention steps that could have been taken. The first involves any of several personal hygiene measures. More specifically these are proper handwashing and scrubbing under fingernails before handling food; avoiding the transfer of saliva from the mouth to food by eating, talking, or chewing gum; and not contaminating hands by touching hair, skin, or nose. The second preventive measure is to ensure that equipment is clean and sanitary before using. The third and most important preventive measure is making certain that the cheddar cheese is kept in the refrigerator at 41° F (5° C) or below rather than at room temperature. Had the cheese been held at temperatures outside the temperature danger zone, *Staphylococcus aureus* could not have grown and produced the heat stable toxin.

In this case study, cooking could not be used as a preventative measure.

Case Study 2.2

The germ, *Clostridium botulinum*, comes from the soil and can be found on just about any food. However, it is usually in its spore, or dormant form, to protect itself from the deadly oxygen in the air. Although the potatoes were washed and scrubbed before baking, some *botulinum* spores still remained on the skins. (You can never totally get rid of microscopic bacterial spores by washing!) The cooking process heat-activated the spores. This means that when the potatoes had been given enough time to cool off, the spores turned into their vegetative, or non-dormant, forms and started reproducing.

To test for doneness, the cooks poked forks into the potatoes. That action carried some of the spores from the skins into the moist, oxygen-free insides of the potatoes. When you cook food, you drive out the oxygen from the food mass. Also, the heat of cooking cannot kill bacterial spores. The spores could then turn vegetative, start reproducing, and release toxin as the germs grew.

In all the years that Restaurant A saved leftover potatoes to make salad, *C. botulinum* had not turned vegetative. That night in 1976, it did. All it took to destroy the reputation of the food establishment was that one incident. This tragedy did not have to happen if they had only known how to handle the leftover potatoes.

Quiz 2.1

1. d	2. a
3. a	4. c
5. b	6. b
7. c	8. b
9. d	10. a

Chapter 3

Case Study 3.1

Due to her illness, Michelle should **not** be handling food under any circumstances. Michelle could easily contaminate the food by coughing, sneezing, or simply touching the food. The food manager should advise Michelle not to continue working until she has recovered from the illness. Under special circumstances, the food manager could assign Michelle to tasks away from food preparation and service, or utensil wash, rinse, and storage areas.

Case Study 3.2

Cameron did not properly clean and sanitize the food thermometer after measuring the temperature of a raw food (meat) followed by a cooked food (eggs). By not cleaning and sanitizing the thermometer, he could contaminate the eggs with the stem of the thermometer that may contain juices from the raw meat. Instead, Cameron should have cleaned the thermometer and sanitized it in an in-place sanitizing solution.

Case Study 3.3

It is likely the food worker's hands became contaminated with feces when they used the toilet. If the workers did not wash their hands properly before returning to work, the Hepatitis A virus could be transferred from their hands to lettuce and other foods. Lettuce is a ready-to-eat food which will not be heated to destroy the virus.

Quiz 3.1

1. a	2. d
3. b	4. b
5. d	6. d
7. d	8. c
9. a	10. b

Chapter 4

Case Study 4.1

1. The sanitation hazards in the walk-in refrigerator are:

   a. Turkeys are not covered.

   b. Raw products (turkeys) are stored above ready-to-eat food (lettuce).

   c. Containers of food are not properly labeled and dated.

   d. Shelves are lined with aluminum foil which reduces air circulation for cooling.

   e.   The thermometer for the walk-in was located near the
        cooling unit, instead of in the warmest part of the
        storage compartment.

2. Violations a, b, and c could result in contamination and
   spoilage.

3. Cover the turkeys during thawing. Store raw foods below
   containers used for food storage. Remove aluminum foil
   from the shelves in the walk-in. Relocate the thermometer to
   the warmest part of the walk-in.

Case Study 4.2

1. Potentially hazardous foods (refrigerated and frozen) are
   not being stored at proper temperatures upon delivery.
   Because foods are not being inspected at the time of
   delivery, defective products may be accepted that would
   normally be sent back to the supplier.

2. Contact suppliers and request that no deliveries be made
   during peak periods such as between 11 am and 2 pm when
   employees are unable to process them and move to proper
   storage conditions. Make certain employees know how to
   inspect incoming shipments and place them in storage in a
   timely fashion.

Case Study 4.3

1. The outbreak of *E. coli* O157:H7 that you read about in this
   case study had a negative influence on consumer confidence
   in the United States. Children and older adults died from *E.
   coli* O157:H7 infection during this outbreak. Foodborne
   illness can kill, and you can prevent that.

2. Undercooked hamburgers played an important role in this
   outbreak. As a result, regulatory agencies recommend
   ground beef patties be cooked to 155°F (68°C) for 15
   seconds for most consumers and 160°F (71°C) for consumers
   in high risk groups.

Quiz 4.1

1. c	2. b
3. d	4. c
5. a	6. c
7. b	8. c
9. c	10. d
11. c	12. d
13. c	14. a
15. d	

Chapter 5

Case Study 5.1

Ingredients	Amount	Servings		
		25	**50**	**100**
Hoisin sauce	Tbsp.	2.5	5.0	10
Bean sauce	Tbsp.	2.5	5.0	10
Apple sauce	Tbsp.	2.5	5.0	10
Catsup	Tbsp.	5.0	10.0	20
Soy sauce	Tbsp.	2.5	5.0	10
Rice wine	Tbsp.	5.0	10.0	20
Peanut oil	Tbsp.	2.5	5.0	10
Ginger, minced	tsp.	1.25	2.5	5
Scallion, minced	1	1.5	5.0	10
Garlic cloves, minced	2	5.0	10.0	20
Sugar	Tbsp.	2.5	5.0	10
Salt	tsp.	1.25	2.5	5
Pork spareribs, trimmed	1 rack	2.6	5.0	10
Honey	Tbsp.	7.5	5.0	30

Pre-preparation

SOP 1   1. Thaw spareribs under refrigeration at 41°F(5°C) for 1 day or until completely thawed.

SOP 2   2. Wash hands before starting food preparation.

3. Combine all of the ingredients for the marinade in a large bowl.

Preparation

SOP 3   4. Score the ribs with a sharp knife. Place the ribs in the marinade. Marinate them for at least 4 hours at a product temperature of 41°F (5°C) or lower in refrigerated storage.

5. Place the spareribs in a smoker at 425°F (220°C) for 30 minutes.

CCP 1   6. Reduce the heat to 375°F (190°C) and continue smoking the ribs to a final internal temperature of 145°F (63°C) or higher. This phase of the cooking process should take about 50 minutes. Brush the honey on the spareribs during the last 5 minutes of the smoking process.

Holding/Service

CCP 2   7. Slice the ribs between the bones and maintain at 140°F (60°C) or above while serving.

Cooling

CCP 3   8. Transfer any unused product into clean, 2-inch deep pans. Quick-chill the product from 140°F (60°C) to 70°F (21°C) within 2 hours and from 70°F (21°C) to 41°F (5°C) within 4 hours.

Storage

CCP 4   9. Store the chilled product in a covered container that is properly dated and labeled. Refrigerate at 41°F (5°C) or below.

Reheating

CCP 5   10. Heat spareribs to an internal temperature of 165°F (74°C); or 190°F (88°C) in a microwave oven within 2 hours.

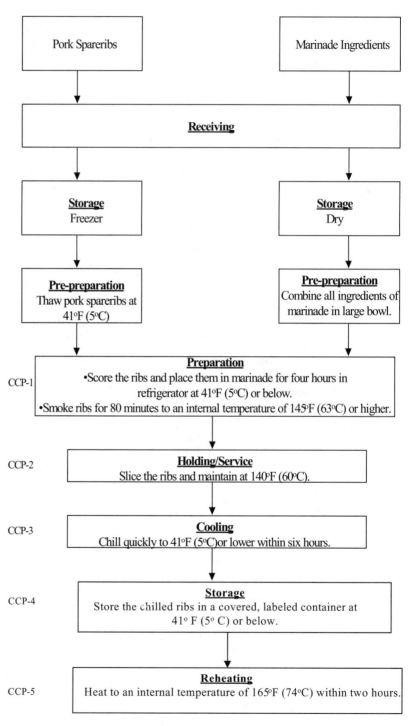

	Pork Spareribs		Marinade Ingredients

Receiving

Storage Freezer		**Storage** Dry

Pre-preparation Thaw pork spareribs at 41°F (5°C)	**Pre-preparation** Combine all ingredients of marinade in large bowl.

CCP-1   **Preparation**
• Score the ribs and place them in marinade for four hours in refrigerator at 41°F (5°C) or below.
• Smoke ribs for 80 minutes to an internal temperature of 145°F (63°C) or higher.

CCP-2   **Holding/Service**
Slice the ribs and maintain at 140°F (60°C).

CCP-3   **Cooling**
Chill quickly to 41°F (5°C) or lower within six hours.

CCP-4   **Storage**
Store the chilled ribs in a covered, labeled container at 41° F (5° C) or below.

CCP-5   **Reheating**
Heat to an internal temperature of 165°F (74°C) within two hours.

Flow Diagram for Pork Spareribs

Case Study 5.2

If the establishment had implemented a HACCP food safety system, it could have avoided situations like the one described in the case study. The HACCP recipe and flow chart would alert food workers to potential hazards and critical control points that can be used to prevent, minimize, or eliminate them. Since the lettuce is an uncooked food, purchasing it from an approved source would be a critical control point. An even more effective critical control point would be a "no bare hands" policy whereby food workers would be instructed not to touch lettuce and other uncooked foods with their bare hands. Instead, they would use spoons, tongs, or other utensils. A standard operating procedure that should be added to your HACCP recipe for the Caesar salad would be for food workers to wash their hands before starting to handle uncooked ready-to-eat foods. Hands should be washed whenever they become contaminated and when employees return from a break or the toilet. The proper use of utensils and the **when and how** of proper handwashing must be taught to food workers as part of your HACCP educational program.

Quiz 5.1

1. a	2. d
3. a	4. b
5. d	6. d
7. b	8. c
9. b	10. b

Chapter 6

Case Study 6.1

1. Wooden cutting boards may harbor disease-causing microbes. Meat is allowed to cool and stay in temperature danger zone. Wiping knife on damp towel transfers microbes on towel onto the blade which then touches the meat.

2. Use approved heat lamps to keep food out of temperature danger zone. Do not use wet kitchen towel to clean equipment.

3. Replace wooden cutting board with board constructed of hard rubber or other approved materials.

Case Study 6.2

1. Contact equipment supplier and get specifications and a price list for different slicers. Compare features and decide which is the best equipment.

2. Does the slicer meet the need and demands of the operation?

   ▼ Can the slicer be used for a long period of time?
   ▼ Does the slicer require extraordinary repair and upkeep?
   ▼ Can the slicer meet future needs of the operation?
   ▼ Cost?

Case Study 6.3

The Plumbing hazard described in this case study is referred to as backsiphonage. In this case, investigators speculate that someone wanted to be neat and rather than curl the tube on the floor, it was tucked into the grate of the drain. this permitted back siphonage to occur and contaminated the ice with raw sewage. This situation could have been prevented by making certain an air gap was provided on the drain line. Plumbing codes typically require a 2-inch air gap between the end of a drain tube and the floor drain. (See Chapter 8.)

Quiz 6.1

1. c	2. a
3. d	4. d
5. c	6. b
7. d	8. a
9. b	10. d

Chapter 7

Case Study 7.1

1. Yes, Ted should be concerned about the suds in the sanitizing solution. Chlorine sanitizers do not work effectively in alkaline conditions. Most detergents are alkaline and, therefore, the effectiveness of the sanitizing solution at the Regal supermarket may be lessened by the suds.

2. Drain the sanitizer and refill the compartment of the sink with fresh water and sanitizer measured at 50 ppm.

3. Make certain the detergent is removed by the fresh water rinse that occurs in the second compartment of the sink.

Case Study 7.2

1. Use the three-compartment sink to manually wash and sanitize equipment and utensils.

2. The facility could use single-service disposable dishware and tableware until the dishwashing machine can be repaired.

3. It is not necessary to suspend food operations provided the measures described in answers 1 and 2 can be provided.

Case Study 7.3

1. When inspectors from the health department removed the housing, they found meat residue. Investigators speculate improper cleaning resulted in ongoing contamination of the ground beef over several days. A stool specimen from one of the butchers, who had eaten the raw ground beef, was positive for *Salmonella typhimurium*.

2. Equipment, such as the meat grinder in this example, must be cleaned and sanitized whenever it becomes contaminated. This should occur at least once a day and more often, if necessary, to control contamination. Some types of food processing equipment, such as a meat grinder, cannot be cleaned and sanitized in a three-compartment sink or a dishwashing machine. Instead, they are designed to be cleaned in place. This type of equipment must be easily

disassembled to provide access to all food contact surfaces. The parts that are removed are cleaned and sanitized in a three-compartment sink or a dishwasher. The remaining food contract surfaces are commonly cleaned and sanitized using a three-bucket technique for washing, rising, and sanitizing. Do not forget to resanitize all food contact surfaces that have been touched while reassembling the equipment.

Quiz 7.1 (Multiple Choice)

1. c	2. b
3. a	4. a
5. c	6. a
7. d	8. c

Chapter 8

Case Study 8.1

1. Floors *and* walls should be nonabsorbent, easily cleanable, durable, and smooth. Walls should be light in color.

2. Wood is porous and absorbs water and other liquids. It would not be durable and easy-to-clean.

3. Any light color would be good. A shade of white is often preferred. Light colors show soil and help food workers determine when a surface has been properly cleaned.

4. Grease, oil, blood, and other soils are common in meat departments. It is important to select floor and wall materials that are nonabsorbent and can be effectively cleaned with soap and water. These materials should also stand up under spray type cleaning procedures.

Case Study 8.2

1. The lack of a properly designed and operating handwashing lavatory in the kitchen at Ryan's may pose a health risk for customers. When food workers cannot wash their hands

when they become contaminated, risk of food contamination and cross contamination is high. Food workers are more likely to wash their hands correctly when there is a convenient sink and supplies near their work station. By having to walk to customer rest rooms, they may be encouraged to wipe their hands on a towel or apron rather then take the time to wash properly.

2. The health department should direct Ryan's to have the lavatory repaired immediately. Until then, food workers should use utensils and disposable gloves to prevent hand contact with food. The infected worker should stay away from work until a physician certifies him or her free of the Hepatitis virus.

3. Train employees about proper handwashing and importance of avoiding hand contact with food. Management should also ask food workers to notify them when they have symptoms of a communicable disease that might be transmitted to food.

4. Ryan's will likely receive poor publicity as a result of the outbreak. This will cause consumer confidence to drop and business will likely decline. The short- and long-term consequences may ruin the restaurant's reputation and cause financial ruin.

Case Study 8.3

Pest control is an essential part of a food safety program. The first step to controlling cockroaches and other pests is good sanitation. Cockroach control involves keeping the insects out of the establishment and depriving them of food and shelter. Cockroaches are hardy insects that resist pest control measures. This makes an adequate pest control program all the more critical.

Quiz 8.1 (Multiple Choice)

1. d	2. a
3. b	4. d
5. b	6. a
7. a	8. b
9. a	10. d

Chapter 9

Case Study 9.1

1. a. Assist guests and employees to safely evacuate the building.

    b. Call emergency medical and police services to aid victims that have suffered injuries due to the earthquake.

    c. Disconnect gas and electricity to avoid fires and explosions.

    d. To the extent possible, ensure the structural safety of the building.

    e. Dispose of all foods which are contaminated with glass and other physical hazards.

2. Fire, police, ambulance, hospital and other types of emergency services. Public and private utilities including, but not limited to, electric, gas, and water companies.

3. In extreme emergency conditions such as the one described in the case study, it is very likely that all food would have to be discarded because of contamination from glass and other physical agents.

    It is extremely difficult to serve safe food without an adequate supply of energy and a safe water supply. If the interruption of the power and water service is only temporary, it might be possible to get by with single-service articles (disposable plates, etc.) and temporary sources of heating and cooling for a brief period of time. Consult with your local regulatory agency before attempting to operate under these less than ideal conditions.

Case Study 9.2

1. If the new employee had been taught the importance of cleaning up spills immediately, the cook might not have fallen.

2. Assess the level of injury to the employee and call for help if needed. Do not make the injury worse by trying to move the victim if he or she appears to be seriously injured.

3. The risk of falls could be reduced by putting anti-slip mats around the fryer. The design and material of these mats should be approved by the local health department. Anyone who falls should remain on the floor until someone checks for injuiry.

Quiz 9.1

1. Mesh gloves

2. Pressure

3. Closed, labeled

4. Remove

5. Third-degree

6. Fire extinguishers

7. Hospitals, fire departments, emergency medical systems, insurance companies, professional associations, Red Cross, and vendors who supply food supplies and equipment

8. Lie still

9. Heimlich Maneuver

10. Emergency phone numbers

Chapter 10

Case Study 10.1

In reference to the situation presented in this case study, did you agree with what the manager told the owner? He said, "We cannot afford the loss of business and reputation, let alone individual problems from contaminated food." Employees who are not trained to handle fryers, slicers, hot and cold holding areas, and food safety and sanitation problems can harm themselves, the equipment, and consumers.

Quiz 10.1

1. c	2. d
3. a	4. b
5. c	6. b
7. a	8. c

Chapter 11

Case Study 11.1

The first thing the waiter should do is be compassionate and understanding. Immediately afterwards, the waiter should contact the food manager—it is the responsibility of the food manager to handle claims of foodborne illness. The food manager should then ensure that the chicken salad is not served to any other customers. The chicken salad should be stored in a separate container in the refrigerator and not be discarded. It is important that the chicken salad be retained so that a health inspector can take a portion to perform tests if necessary. Remember to keep some of the chicken salad in case you decide to perform tests with an independent laboratory. The food manager should then contact the local health department and proceed with the investigation. Once the health inspector and/or other regularity agencies become involved, it is important that the food managers cooperate in any way possible. It may be important to provide records of food suppliers and of handling/preparation practices used regarding the chicken salad.

Case Study 11.2

The inability to sanitize equipment and utensils is a very serious situation for a food establishment. During the post-inspection conference. Jon should discuss his options with Inspector Jones. The only soultion to the problem is to have a service person come to the establishment and fix the dispenser as soon as possible. In the meantime, Jon should have his staff wash and sanitize the equipment and utensils using the three-compartment sink. Jon might reduce the volume of dishes and utensils by using plastic utensils and paper cups. Jon should notify Inspector Jones when the sanitizer dispenser has been fixed.

Quiz 11.1

1. b	2. c
3. d	4. c
5. d	6. a
7. a	8. d
9. c	10. d

APPENDIX B

Summary of Agents That Cause Foodborne Illness

CAUSATIVE AGENT (*Sporeforming bacteria)	TYPE OF ILLNESS	SYMPTOMS ONSET	COMMON FOODS	PREVENTION
Anasakis spp,	Parasitic infection	Coughing, vomiting 1 hour to 2 weeks	Raw or undercooked seafood - especially bottom-feeding fish	Cook fish to the proper temperature throughout; freeze to meet *Food Code* specification
Bacillus cereus	Bacterial intoxication or toxin-mediated infection	1) Diarrhea, abdominal cramps (8-16 hrs) 2) Vomiting type- vomiting, diarrhea, abdominal cramps (30 minutes - 6 hrs)	1) Diarrhea type: meats, milk, vegetables 2) Vomiting type: rice, starchy foods; grains and cereals	Properly heat, cool, and reheat foods
Campylobacter jejuni	Bacterial infection	Watery, bloody diarrhea (2-5 days)	Raw chicken, raw milk, raw meat	Properly handle and cook foods; avoid cross contamination
Ciguatoxin	Fish toxin, originating from toxic algae of tropical waters	Vertigo, hot/cold flashes, diarrhea, vomiting (15 minutes - 24 hours)	Marine finfish, including groupers, barracuda, snappers, jacks, mackerel, triggerfish, reef fish	Purchase fish from a reputable supplier; cooking WILL NOT inactivate the toxin

CAUSATIVE AGENT (*Sporeforming bacteria)	TYPE OF ILLNESS	SYMPTOMS ONSET	COMMON FOODS	PREVENTION
Clostridium botulinum	Bacterial intoxication	Dizziness, double vision, difficulty in breathing and swallowing, headache (12-36 hours)	Improperly canned foods, vacuum packed refrigerated foods; cooked foods in anaerobic mass	Properly heat process anaerobically packed foods; DO NOT use home canned foods
Clostridium perfringens	Bacterial toxin-mediated infection	Intense abdominal pains and severe diarrhea (8-22 hours)	Spices, gravy, improperly cooled foods (especially meats and gravy dishes)	Properly cook, cool, and reheat foods
Cryptosporidium parvum	Parasitic infection	Severe watery diarrhea within 1 week of ingestion	Contaminated water; food contaminated by infected food workers	Use potable water supply; practice good personal hygiene and handwashing
Cyclospora cayetanensis	Parasitic infection	Watery and explosive diarrhea, loss of appetite, bloating (1 week)	Water, strawberries, raspberries, and raw vegetables	Good sanitation reputable supplier
Escherichia coli O157:H7	Bacterial infection or toxin-mediated infection	Bloody diarrhea followed by kidney failure and hemolytic uremic syndrome (HUS) in severe cases (12-72 hours)	Undercooked hamburger, raw milk, unpastuerized apple cider, and lettuce	Practice good food sanitation, handwashing; properly handle and cook foods
Giardia lamblia	Parasitic infection	Diarrhea within 1 week of contact	Contaminated water	Potable water supply; good personal hygiene and handwashing
Hepatitis A	Viral infection	Fever, nausea, vomiting, abdominal pain, fatigue, swelling of the liver, jaundice (15-50 days)	Foods that are prepared with human contact; contaminated water	Wash hands and practice good personal hygiene; avoid raw seafood

CAUSATIVE AGENT (*Sporeforming bacteria)	TYPE OF ILLNESS	SYMPTOMS ONSET	COMMON FOODS	PREVENTION
Listeria monocytogenes	Bacterial infection	1) Healthy adult: flu-like symptoms 2) At risk population: septicemia, meningitis, encephalitis birth defects (1 day-3 weeks)	Raw milk, dairy items, raw meats, refrigerated ready-to-eat foods, processed ready-to-eat meats such as hot dogs, raw vegetables, and seafood	Properly store and cook foods, avoid cross contamination; rotate processed refrigerated foods using FIFO to ensure timely use
Mycotoxins	Intoxication	1) Acute onset: hemorrhage, fluid buildup, possible death. 2) Chronic: cancer from small doses over time	Moldy grains: corn, corn products, peanuts, pecans, walnuts, and milk	Purchase food from a reputable supplier; keep grains and nuts dry; and protect products from humidity
Norwalk virus	Viral infection	Vomiting, diarrhea, abdominal pain, headache, and low grade fever; onset 24-48 hrs	Sewage contaminated water; contaminated salad ingredients, raw clams, oysters	Use potable water; cook all shellfish; handle food properly; meet time temperature guidelines for PHF
Rotavirus	Viral infection	Diarrhea (especially in infants and children), vomiting, low grade fever; 1-3 days onset; lasts 4-8 days	Sewage contaminated water; contaminated salad ingredients, raw seafood	Good personal hygiene and handwashing; proper food handling practices
Salmonella spp.	Bacterial infection	Nausea, fever, vomiting, abdominal cramps, diarrhea (6-48 hours)	Raw meats: raw poultry, eggs, milk, dairy products	Properly cook foods, avoid cross contamination

CAUSATIVE AGENT (*Sporeforming bacteria)	TYPE OF ILLNESS	SYMPTOMS ONSET	COMMON FOODS	PREVENTION
Scombrotoxin	Seafood toxin originating from histamine producing bacteria	Dizziness, burning feeling in the mouth, facial rash or hives, peppery taste in mouth; headache, itching, teary eyes runny nose (1 -30 minutes)	Tuna, mahi-mahi, bluefish, sardines, mackerel, anchovies, amberjack, abalone	Purchase fish from a reputable supplier, store fish at low temperatures to prevent growth of histamine-producing bacteria, toxin IS NOT inactivated by cooking
Shellfish toxins: PSP, DSP, DAP, NSP	Intoxication	Numbness of lips, tongue, arms, legs, neck; lack of muscle coordination (10 to 60 minutes)	Contaminated mussels, clams, oysters, scallops	Purchase from a reputable supplier
Shigella spp.	Bacterial infection	Bacillary dysentery, diarrhea, fever, abdominal cramps, dehydration (1-7 days)	Foods that are prepared with human contact: salads, raw vegetables, milk, dairy products, raw poultry, non-potable water, ready-to-eat meat	Wash hands and practice good personal hygiene; properly cook foods
Staphylococcus aureus	Bacterial intoxication	Nausea, vomiting, abdominal cramps, headaches (2-6 hrs)	Foods that are prepared with human contact, cooked or processed foods	Wash hands and practice good personal hygiene. Cooking WILL NOT inactivate the toxin

CAUSATIVE AGENT (*Sporeforming bacteria)	TYPE OF ILLNESS	SYMPTOMS ONSET	COMMON FOODS	PREVENTION
Toxoplasma gondii	Parasitic infection	Mild cases of the disease involves swollen lymph glands, fever, headache, and muscle aches. Severe cases may result in damage to the eye or the brain. (10-13 days)	Raw meats; raw vegetables and fruit	Good sanitation, reputable supplier, proper cooking
Trichinella spiralis	Parasitic infection from a nematode worm	Nausea, vomiting, diarrhea, sweating, muscle soreness (2-28 days)	Primarily undercooked pork products and wild game meats (bear, walrus)	Cook foods to the proper temperature throughout
Vibrio spp.	Bacterial infection	Headache, fever, chills, diarrhea, vomiting, severe electrolyte loss, gastroenteritis (2-48 hrs)	Raw or improperly cooked fish and shellfish	Practice good sanitation, properly cook foods, avoid serving raw seafood

APPENDIX C

Conversion Table for Fahrenheit and Celsius for Common Temperatures Used in Food Establishments

°F	°C		°F	°C
212	100		100	38
200	93		75	24
194	90		70	21
190	88		55	13
180	82		45	7
171	77		41	5
165	74		38	3
160	71		36	2
155	68		33	1
150	66		32	0
145	63		30	-1
140	60		29	-2
130	54		0	-18
120	49		-4	-20
110	43		-31	-35

APPENDIX D

Tools for Designing and Implementing HACCP Programs

Ingredients

▲ Does the food contain any sensitive ingredients (e.g., raw meat, poultry, fish, shellfish, eggs, or milk) that are likely to introduce microbiological, chemical, or physical hazards?

▲ Has this food been implicated in previous outbreaks? How frequently?

Intrinsic Factors of Food

▲ Which physical characteristics and composition of the food (e.g., pH, water activity, and preservatives) must be controlled in order to ensure food safety?

▲ Does the food permit survival or multiplication of infectious and/or toxin-producing microorganisms in the food before or during preparation?

▲ Will the food permit survival or multiplication of infectious and/or toxin-producing microorganisms during later stages of preparation, storage, or consumer possession?

▲ Are there other similar products in the market place? What has been the safety record for these products?

Procedures Used for Preparation and Processing

▲ Does the process used to prepare or process the food include a manageable step that will destroy infectious microorganisms or prevent toxin formation?

Figure D.I—Questions Included in a Hazard Analysis Checklist

▲ Is the product subject to recontamination between preparation and service or packaging?

Microbial Content of the Food

▲ Is the food commercially sterile? Is it a low acid (pH of 4.6 or above) canned item that has been heat treated to destroy both vegetative cells and spores of foodborne disease-causing microorganisms?

▲ Is it likely that the food will contain living microorganisms that are capable of causing foodborne illness?

▲ What is the normal microbial content of the food stored under proper conditions?

▲ Does the microbial population alter the safety of the food?

Facility Design

▲ Does the layout of the kitchen adequately separate raw foods from ready-to-eat foods to prevent cross contamination?

▲ Do the traffic patterns of workers and food flow present a potential source of contamination?

Equipment Design

▲ Will the kitchen equipment provide the time and temperature needed to ensure safe food?

▲ Is the size of the kitchen equipment adequate for the volume of food that will be prepared?

▲ Is the kitchen equipment reliable or does it break down frequently?

▲ Is there a chance for product contamination with hazardous substances?

▲ Is kitchen equipment equipped with product safety devices, such as time/temperature controls, to boost food safety?

Packaging

▲ Does the type of food package support the growth of disease-causing germs and/or the formation of toxins?

Figure D.I—Questions Included in a Hazard Analysis Checklist (Cont.)

▲ Is the food's packaging material resistant to damage, thereby preventing the entrance of microbial contamination?

▲ Is the package clearly labeled "Keep Refrigerated" if this is required to preserve and protect the food?

▲ Does the package include adequate instructions for the safe handling and preparation of the food by the consumer?

▲ Are tamper-evident packaging features used?

▲ Is each package legibly and accurately coded to indicate production lot?

▲ Does each package have a label that contains the required information?

Sanitation

▲ Can the sanitation practices that are employed in the food establishment impact upon the safety of the food that is being prepared?

▲ Can the facility be properly cleaned and sanitized to permit the safe handling of food?

▲ Is it possible to consistently provide sanitary conditions in the kitchen to ensure production of safe foods?

Employee Health, Hygiene, and Education

▲ Can employee health or personal hygiene practices affect the safety of the food being prepared?

▲ Do food workers understand the food preparation process and the factors that they must control to ensure safe foods?

▲ Will food workers inform managers of possible problems which could affect food safety when they suspect them?

Conditions of Storage Between Packaging and the Consumer

▲ Is it likely that food will be improperly stored at the wrong temperature?

▲ Would storage at improper temperatures lead to unsafe food?

Figure D.I—Questions Included in a Hazard Analysis Checklist (Cont.)

Intended Use

▲ Is the food intended for consumption by people who do not have an increased risk of becoming ill?

▲ Is the food intended for consumption by high risk populations such as very young children, frail elderly, pregnant women and those with suppressed immune systems?

Figure D.I—Questions Included in a Hazard Analysis Checklist (Cont.)

Critical Control Point Decision Tree

Consider conditions unique to your establishment when developing a HACCP plan for your facility. The decision tree for determining critical control points presented in the following **Critical Control Point Decision Tree** may prove helpful in verifying which of the steps in your food flow are CCPs.

The Modified CCP Decision Tree

Q1. Does this step involve a hazard of sufficient risk and severity to warrant its control?

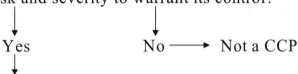

Yes   No ⟶ Not a CCP

Q2. Does a preventive measure for the hazard exist at this step?

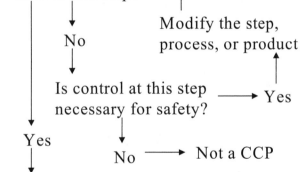

Modify the step, process, or product

No

Is control at this step necessary for safety? ⟶ Yes

Yes

No ⟶ Not a CCP

Q3. Is control at this step necessary to prevent, eliminate, or reduce the risk of the hazard to consumers?

Yes ⟶ CCP   No ⟶ Not a CCP

Source: *HACCP-Establishing Hazard Analysis Critical Control Point Programs: A Workshop Manual.* The Food Processors Institute, Washington, D.C. 1995.

Figure D.2—Critical Control Point Decision Tree

APPENDIX E

Specific Elements of Knowledge That Every Retail Food Protection Manager Should Know

I. Identify terms related to foodborne illness.

    A. Define terms associated with foodborne illness.

        1. foodborne illness

        2. foodborne outbreak

        3. foodborne infection

        4. foodborne intoxication

        5. disease communicated by food

        6. foodborne pathogens

    B. Recognize the major microorganisms and toxins that can contaminate food and the problems that can be associated with the contamination.

        1. bacteria

        2. viruses

        3. parasites

        4. fungi

    C. Define and recognize potentially hazardous foods.

    D. Define and recognize chemical and physical contamination and associated illness.

    E. Define and recognize the major contributing factors for foodborne illness.

    F. Recognize how microorganisms cause foodborne illness.

II. Identify time/temperature relationships with food.

A. Recognize the relationship between time/temperature and microorganisms (survival, growth, and toxin production) during the following stages:

1. receiving

2. storing

3. thawing

4. cooking

5. holding/displaying

6. serving

7. cooling

8. storing (post-production)

9. reheating

10. transporting

B. Describe the use of thermometers in monitoring food temperatures.

1. types of thermometers

2. techniques and frequency

3. calibration and frequency

III. Describe the relationship between personal hygiene and food safety.

A. Recognize the association of hand contact and foodborne illness.

1. hand washing—technique and frequency

2. proper use of gloves, including how often to replace

3. minimal hand contact with food

B. Recognize the association of personal habits and behaviors and foodborne illnesses.

1. smoking

2. eating and drinking

3. wearing clothing that may contaminate food

    4. personal behaviors, including coughing, sneezing, etc.

C. Recognize the association of the health of the food handler to foodborne illness.

    1. free from communicable disease symptoms

    2. free of infections that can be spread through contact with food

    3. protect food from contact with open wounds

D. Recognize how policies, procedures, and management can improve food safety and hygiene practices.

IV. Describe methods for preventing food contamination from purchasing to serving.

A. Define terms associated with contamination:

    1. contamination

    2. adulteration

    3. damage

    4. approved source

    5. sound and safe conditions

B. Identify potential hazards and methods prior to and during delivery.

    1. approved source

    2. sound and safe condition

C. Identify potential hazards and methods to minimize or eliminate hazards after delivery.

    1. personal hygiene

    2. cross contamination

        a. food to food

        b. equipment and utensils

    3. contamination

        a. chemical

        b. additives

        c. physical

    4. service/display—customer contamination

    5. storage

    6. re-service

V. Identify and apply correct procedures for cleaning and sanitizing.

  A. Define terms associated with cleaning and sanitizing.

    1. cleaning

    2. sanitizing

  B. Apply principles of cleaning and sanitizing.

  C. Identify material: equipment, detergent, sanitizer.

  D. Apply appropriate methods of cleaning and sanitizing.

    1. manual warewashing

    2. mechanical warewashing

    3. clean-in-place

  E. Identify frequency of cleaning and sanitizing.

VI. Recognize problems and potential solutions associated with facility, equipment, and layout.

  A. Identify facility, design, and construction suitable for food establishments:

    1. refrigeration

    2. heating and cooling

    3. floors, walls, and ceilings

    4. pest control

    5. lighting

    6. plumbing

    7. ventilation

    8. water supply

    9. wastewater disposal

    10. waste disposal

B. Recognize problems and potential solutions associated with temperature control, preventing cross contamination, housekeeping, and maintenance. Implement:

1. self-inspection program

2. pest control program

3. cleaning schedules and procedures

4. equipment and facility maintenance program

(Developed by the Conference for Food Protection)

(Source: *Proceedings of the Conference for Food Protection Meeting, April 10-15, 1996, Denver, CO, pp. 75-79.*)

GLOSSARY

Abrasive cleaners - cleaning compound containing finely ground minerals used to scour articles; can scratch surfaces.

Acceptable level - within an established range of safety.

Accessible - easy to reach or enter.

Acid - a substance with a pH of less than 7.0.

Additives - natural and man-made substances added to a food for an intended purpose (such as preservatives and colors) or unintentionally (such as pesticides and lubricants).

Adulterated - the deliberate addition of inferior or cheaper material to a supposedly pure food product in order to stretch out supplies and increase profits.

Aerobe - an organism, especially a bacterium, that requires oxygen to live.

Air curtain - a window or doorway equipped with jets that force air out and across the opening to keep flying insects from entering.

Air-dry - to dry in room air after cleaning, washing, or sanitizing.

Air gap - an unobstructed, open vertical distance through air that separates an outlet of the potable water supply from a potentially contaminated source like a drain.

Alkaline - a substance that has a pH of more than 7.0.

Anaerobe - an organism, especially a bacterium, that does not require oxygen or free oxygen to live.

Anisakiasis **-** disease caused by the anisakis parasite.

Anisakis - roundworm parasite found in fish.

Aseptic packaging - a method in which food is sterilized or commercially sterilized outside the can and then aseptically placed in previously sterilized containers which are then sealed in an aseptic environment. This method may be used for liquid foods such as concentrated milk and soups.

At risk - the term used to describe individuals such as infants, children, pregnant women, and those with weakened immune systems for whom foodborne illnesses can be very severe, even life threatening.

Backflow - flow of contaminated water into potable supply; caused by back pressure.

Backsiphonage - a form of backflow that can occur when pressure in the potable water supply drops below pressure in the flow of contaminated water.

Bacteria - single-celled microscopic organisms.

Bactericide - substance that kills bacteria.

Bacterium - one microorganism.

Bi-metallic stemmed thermometer - food thermometer used to measure product temperatures.

Binary fission - the process by which bacteria grow. One cell divides to form two new cells.

Biofilm - a layer of stubborn soil that develops on surfaces that are improperly cleaned or sanitized; bacteria can accumulate and grow on this surface.

Blast chiller - special refrigerated unit that quickly freezes food items.

Botulism - type of food intoxication caused by *C. botulinum*.

Calibrate - to determine and verify the scale of a measuring instrument with a standard. Thermometers used in food establishments are commonly calibrated using an ice slush method (32°F or 0°C) or a boiling point method (212°F or 100°C).

Campylobacter jejuni - a microaerophilic nonsporeforming bacterium that causes a foodborne illness.

Cardiopulmonary resuscitation (CPR) - first aid technique to apply to persons who have stopped breathing.

Chemical sanitizers - products used on equipment and utensils after washing and rinsing to reduce the number of disease-causing microbes to safe levels.

Ciguatoxin - a toxin from reef feeding fish that causes foodborne illness.

Clean - free of visible soil but not necessarily sanitized; surface must be clean before it can be sanitized.

Cleaning - removal of visible soil but not necessarily sanitized. A surface must be cleaned before it can be sanitized.

Cleaning agent - a chemical compound formulated to remove soil and dirt.

Clean-in-Place (CIP) -equipment designed to be cleaned without moving, usually large or very heavy.

Code - a systematic collection of regulations, or statutes, and procedures designed to protect the public.

Cold-holding - refers to the safe temperature range (less than 41°F or 5°C) for maintaining foods cold prior to service for consumption.

Competency based training - a job is analyzed and broken down into steps which must be learned to gain mastery of the task; training should focus on those steps.

Contamination - the unintended presence of harmful substances or conditions in food that can cause illness or injury to people who eat the infected food.

Cooking - the act of providing sufficient heat and time to a given food to effect a change in food texture, aroma, and appearance. More importantly, cooking ensures the destruction of foodborne pathogens inherent to that food.

Cooling - the act of reducing the temperature of properly cooked food to 41°F (5°C) or below.

Corrosion resistant - materials which maintain their original surface characteristics under continuous use in food service with normal use of cleaning compounds and sanitizing solutions.

Coving - a curved sealed edge between the floor and wall that makes cleaning easier and inhibits insect harborage.

Critical Control Point (CCP) - means a point or procedure in a specific food system where loss of control may result in an unacceptable health risk.

Critical limit - the maximum or minimum value to which a physical, biological, or chemical parameter must be controlled at a critical control point to minimize the risk that the identified food safety hazard may occur.

Cross-connection - any physical link through which contaminants from drains, sewers, or waste pipes can enter a potable water supply.

Cross contamination - transfer of harmful organisms between items.

Cumulative - increasing in effect by successive additions. For example, hot food must be cooled from 140°F (60°C) to 70°F (21°C) within 2 hours; it must reach 41°F (5°C) within an additional 4 hours to prevent bacterial growth. Therefore, the *cumulative time the food is in this danger zone equals a total of 6 hours or less.*

Danger zone - temperatures between 41°F (5°C) and 140°F (60°C).

Disinfectant - destroys harmful bacteria.

Dead man's switch - activates equipment when depressed and stops if pressure is relieved.

Detergents - cleaning agent which contains surfactants used with water to break down soil to make it easier to remove.

Deviation - to diverge; to go in different directions.

Digital thermometer - a battery-powered thermometer that reveals temperature in a digital numerical display.

Easy-to-clean - materials and design which facilitate cleaning.

Easy-to-move - on wheels, raised on legs, or otherwise designed to facilitate cleaning.

Employee - person working in or for a food establishment who engages in food preparation, service, or other assigned activity.

Equipment - the appliances: stoves, ovens, etc.; and storage, such as refrigerated units used in food establishments.

Evaluation procedures - systematically checking progress to determine if goals have been met.

Facultative anaerobe - an organism that can grow with or without free oxygen.

FATTOM - an acronym used to indicate the six conditions bacteria need for growth. These conditions are **Food**, **Acid**, **Temperature**, **Time**, **Oxygen**, and **Moisture**.

FIFO - acronym for *first in, first out,* used to describe stock rotation procedures of using older products first.

Food establishment - an operation that stores, prepares, packages, serves, vends, or otherwise provides food for human consumption such as a restaurant, food market, institutional feeding location, or vending location.

Foodborne disease outbreak - an incident in which two or more people experience a similar illness after ingesting a common food, and epidemiological analysis identifies the food as the source of the illness.

Foodborne illness - an illness caused by the consumption of a contaminated food.

Food-contact surface - any surface of equipment or any utensils which food normally touches.

Footcandle - unit of lighting equal to the illumination one foot from a uniform light source.

Fungi - a group of microorganisms that includes mold and yeasts.

Garbage - wet waste matter, usually food product, that cannot be recycled.

Gastroenteritis - an inflammation of the linings of stomach and intestines which can cause nausea, vomiting, abdominal cramping, and diarrhea.

Germicide - a substance which kills harmful microbes and germs.

Germs - general term for microorganisms, including bacteria and viruses.

Grade standards - primarily standards of quality to help producers, wholesalers, retailers, and consumers in marketing and purchasing food products. The grade standards are not aimed at protecting the health of the consumer but rather at ensuring value received according to uniform quality standards.

GRAS substances - GRAS stands for generally recognized as safe. These are substances added to foods that have been shown to be safe based on a long history of common usage in food.

Hand washing - the proper cleaning of hands with soap and warm water to remove dirt, filth, and disease germs.

Harborage - shelter for pests.

Hazard - means a biological, chemical, or physical agent that may cause an unacceptable consumer health risk.

Hazard analysis - identify hazards (problems) that might be introduced into food by unsafe practices or the intended use of the product.

Hazard Analysis Critical Control Point (HACCP) - a food safety assurance system that highlights potential problems in food preparation and service.

Heimlich maneuver - method used to expel foreign body caught in someone's throat.

Hermetic packaging - a container that is sealed completely against the entry of bacteria, molds, yeasts, and filth as long as it remains intact.

Hot-holding - refers to the safe temperature range of 140°F (60°C) and above to maintain properly cooked foods hot until served.

Immunosuppressed - an individual who is susceptible to illness because his or her immune system is not working properly or is weak.

Impermeable - does not permit passage, especially of fluids.

Infection - illness caused by eating food that contains living disease-causing microorganisms.

Infestation - presence of a large number of pests.

Ingestion - the process of eating and digesting food.

Inherent - being an essential part of something.

Integrated Pest Management (IPO) - a system of preventive and control measures used to control or eliminate pest infestations in food establishments.

Intoxication - Illness caused by eating food that contains a harmful chemical or toxin.

Irradiation - exposure of food to low level radiation to prolong shelf life and eliminate pathogens.

Kitchenware - utensils used to prepare food.

Larva - immature stage of development of insects and parasites.

Leftovers - any food that is prepared for a particular meal and is held over for service at a future meal.

Manager - the individual present at a food service establishment who supervises employees who are responsible for the storage, preparation, display, and service of food to the public.

Measuring device - usually a thermometer that registers the temperature of products and water used for sanitizing.

Media - newspapers, magazines, radio, and television.

Mesophile - microorganisms that grow best at moderate temperatures.

Microbe - a microscopic organism such as a bacterium or a virus.

Microorganism - bacteria, viruses, molds, and other tiny organisms that are too small to be seen with the naked eye. The organisms are also referred to as microbes because they cannot be seen without the aid of a microscope.

Misbranding - falsely or misleadingly packaged or labeled food; may contain ingredients not included on label or does not meet national standards for that food.

Modified Atmosphere Packaging (MAP) - a food processing technique where foods are placed in a flexible container and the air is removed from the package. Gases may be added to help preserve the food.

Mold - any of various fungi that spoils food and has a fuzzy appearance.

Monitoring procedures - a defined method of checking foods during receiving, storage, preparation, holding, and serving processes.

Non food-contact surface - any area not designed to touch food.

Onset time - the period between eating a contaminated food and developing symptoms of a foodborne illness.

Palatable - food that has an acceptable taste and flavor.

Parasite - an animal or plant that lives in or on another and from whose body it obtains nourishment.

Parts per million (ppm) - unit of measure for water hardness and chemical sanitizing solution concentrations.

Pasteurization - a low heat treatment used to destroy disease-causing organisms and/or extend the shelf life of a product by destroying organisms and enzymes that cause spoilage.

Pathogenic - capable of causing disease; harmful; any disease-causing agent.

Perishable - quick to decay or spoil unless stored properly.

Personal hygiene - health habits including bathing, washing hair, wearing clean clothing, and proper hand washing.

Pest control operator - licensed individual or certified technician who provides pest control services.

pH - the symbol that describes the acidity or alkalinity of a substance, such as food.

Physical hazard - particles or fragments of items not supposed to be in foods.

Potable water - water that is safe to drink.

Potentially hazardous food - a food that is natural or man made and is in a form capable of supporting the rapid and progressive growth of infectious and toxin-producing microorganisms. The foods usually have high protein and moisture content and low acidity.

Pounds per square inch (psi) - amount of pressure per square inch.

Premises - physical environment of the food establishment; grounds, interior, and exterior of building(s).

Preventive measures - procedures that keep foods from becoming contaminated.

Psychrophiles - microorganisms that grow best at cold temperatures.

Quaternary ammonium - a chemical sanitizing compound that is relatively safe for skin contact and is generally noncorrosive; effective in both acid and alkaline solutions.

Ready-to-eat foods - foods that do not require cooking or further preparation prior to consumption.

Reconstitute - to combine dehydrated foods with water or other liquids to bring back to original state; example, addition of water to powdered milk.

Refuse - trash, rubbish, waste.

Reheating - the act of providing sufficient heat (at least 165°F or 74°C) within a two-hour time period to ensure the destruction of any foodborne pathogens that may be present in that cooked and cooled food.

Risk - the chance of injury, damage, or loss.

Sanitary - Clean and free of harmful microorganisms and other contaminants.

Sanitation - maintenance of conditions which are clean and promote good health.

Sanitizer - approved substance or method to use when sanitizing.

Sanitizing - application of an agent that reduces microbes to safe levels.

Sealed - closed tightly.

Selectivity - chemical sanitizers, especially quats, may kill only certain organisms and not others.

Sensitive ingredient - an ingredient that is prone to support bacterial growth.

Single-use articles - items for one use and then discarded, such as paper cups or plastic eating utensils.

Silicate - salt or ester derived from silica; a hard, glassy material found in sand.

Sneeze guard - a clear, solid barrier which partially covers food in self-service areas to keep customers from coughing, sneezing, or projecting droplets of saliva directly onto food.

Soil - dirt and filth.

Sous vide - A French term for "under vacuum." A method of food processing that involves placing food ingredients in plastic pouches and vacuuming the air out. The pouch is then minimally cooked under precise conditions and refrigerated immediately.

Splash contact surfaces - areas that are easy to clean and designed to catch splashed substances in work areas.

Spoilage - significant food deterioration, usually caused by bacteria and enzymes, that produces a noticeable change in the taste, odor, or appearance of the product.

Spore - the inactive or dormant state of some rod-shaped bacteria.

Stationary equipment - equipment that is permanently fastened to the floor, table, or countertop,

Sulfites - preservatives used to maintain freshness and color of fresh fruits and vegetables; subject to state regulations.

Surfactant - chemical agent in detergent that reduces the surface tension, allowing the detergent to penetrate and soak soil loose; wetting agent.

Tableware - plates, cups, bowls, etc.

Task analysis - examining a job task to determine what it takes to do the job.

Temperature abuse - allowing foods to remain in the temperature danger zone (41° and 140°F, 5° and 60°C) for an unacceptable period of time.

Temperature danger zone - temperatures between 41° and 140°F (5° and 60°C) at which bacteria grow best.

Test kit - device that accurately measures the concentration of sanitizing solutions to ensure that they are at proper levels.

Thermometer - a device that measures temperatures.

Thermophiles - microorganisms that grow best at hot temperatures.

Toxin - a poisonous substance produced by microorganisms, plants, and animals and which causes various diseases.

Toxin-mediated infection - illness caused by eating a food that contains harmful microorganisms that produce a toxin once they get inside the human intestinal tract.

Vacuum breaker - designed for use under a continuous supply of pressure. Spring loaded device to operate after extended periods of hydrostatic pressure.

Vegetative state - the active state of a bacterium where the cell takes in nourishment, grows, and produces wastes.

Ventilation - air circulation that removes smoke, odors, moisture, and grease-laden vapors from a room and replaces them with fresh air.

Verification - to prove to be true by evidence, usually a record of time and temperatures of food from receiving to serving or vending.

Viruses - any of a group of infectious microorganisms that reproduce only in living cells. They cause diseases such as mumps and Hepatitis A and can be transmitted through food.

Water activity (A_w) - a measure of the free moisture in a food. Pure water has a water activity of 1.0, and potentially hazardous foods have a water activity of 0.85 and higher.

Wetting agent - a substance which breaks down the soil to allow water and soap or detergent to liquefy and remove dirt and grease.

Wholesome - something that is favorable to or promotes health.

Yeast - type of fungus that is not known to cause illness when present in foods but can cause damage to food products and will change taste; useful in making products such as bread and beer.

INDEX

"I'd like to take THE test, please."

By now you have at least glanced at the book *Essentials of Food Safety & Sanitation*. Perhaps you have determined that you are ready to take THE test so that you can become a **certified food safety manager**. Before you sharpen your pencil, there are a couple of additional questions worth thinking about.

First, why do you even need to become a *certified food safety manager*?

▼ You may just love the study and prevention of foodborne illnesses, but certification is probably required by your company, your boss, or to comply with government regulations. Whatever the reason, you need a respected certification to go forward in your career.

Second, what do you get when you successfully complete this exam?

▼ You receive a nationally recognized certification as a **Food Safety Manager**. You have mastered the knowledge needed to successfully manage the safety and sanitation of food establishments.

▼ Additionally, your name will be listed in the **National Registry of Food Safety Professionals**. This will allow your 3-year certification to be verified by contacting the **National Registry of Food Safety Professionals**. This can be a great boost to your career.

"If I could ask anything . . . I'd like to know about the test."

We trust you mean "What is on the test?"

▼ Here is the examination content list and the number of questions for each:

Ensure Food Protection	12
Purchase and Receive Food	11
Store Foods and Supplies	11
Prepare Foods	9
Serve and Display Foods	9
Use and Maintain Tools and Equipment	2
Clean and Sanitize Equipment, Utensils, and Food Contact Surfaces	4
Select, Monitor, and Maintain Water Sources	1
Monitor and Maintain Plumbing Fixtures	1

TOTAL NUMBER OF QUESTIONS: 80

What about the exam mechanics?

▼ The exam includes 80 multiple-choice questions.

▼ The Exam is in easy-to-understand, simple language and requires an 8th grade reading level.

▼ You will have two hours for the exam.

▼ The exam may be taken in "paper and pencil" form under the supervision of a proctor, or by computer in supervised conditions.

▼ The exam is scored and scaled. To pass the exam you must earn a **75** or higher.

▼ After you take the test, you will be sent a full breakout of how you did. If you don't pass the exam you can use this to analyze where you need to work harder to pass the exam the next time.

How should I prepare for the Certified Food Safety Manager Exam?

It depends . . .

▼ Some states or professional affiliations require attendance in class before sitting for the exam. Seek out your local community college or university for a class or short course.

▼ If you are an independent learner, you should read *Essentials of Food Safety & Sanitation* before you request and take the exam.

▼ You might skip preparing and just keep taking the exam until you get it right. It is possible that your professional experience and commitment to preventing foodborne illnesses has equipped you for the exam.

What or who is Professional Testing, Inc.?

▼ Professional Testing, Inc., prepares and administers certification and licensure examinations for national, state, and local regulatory and certification agencies.

▼ Over the past 25 years, Professional Testing, Inc., has prepared and administered the certification and licensure exams for such professions as electrical contractors, physicians, architects, real estate agents and airline pilots, to name a few.

▼ In 1995, Professional Testing, Inc., brought together leaders from industry, regulatory agencies, and academia to form an expert panel of food professionals. All questions appearing on the Certified Food Safety Manager exam are written, reviewed, and tested by these experts. An advisory board provides continuous monitoring of changes in the food industry to ensure that the examination reflects current practices. The advisory board adheres to the developing Standards for Food Protection Manager Certification, prepared by the Conference for Food Protection.

To find out more about the comprehensive *Certified Food Safety Manager* exam just write or call:

Professional Testing, Inc.
National Registry of Food Safety Professionals
1200 E. Hillcrest Street, Suite 300
Orlando Florida 32803-4737

Phone: (800) 446-0257 between 9:00 a.m. – 4:00 p.m. EST
Fax: (407) 894-7748